Acid Rain

SCIENCE, TECHNOLOGY, AND SOCIETY SERIES

(Formerly Monographs on Science, Technology, and Society)

1 Eric Ashby and Mary Anderson *The politics of clean air*
2 Edward Pochin *Nuclear radiation: risks and benefits*
3 L. Rotherham *Research and innovation: a record of the Wolfson Technological Projects Scheme 1968–1981*; with a foreword and post-script by Lord Zuckerman
4 John Sheail *Pesticides and nature conservation 1950–1975*
5 Duncan Davies, Diana Bathurst, and Robin Bathurst *The telling image: the changing balance between pictures and words in a technological age*
6 L.E.J. Roberts, P.S. Liss, and P.A.H. Saunders *Power generation and the environment*
7 Roy Gibson *Space*
8 B.J. Mason *Acid rain: its causes and its effects on inland waters*

Acid Rain
Its Causes and its Effects on Inland Waters

B.J. MASON

(Formerly Director of the Anglo-Scandinavian Surface Waters Acidification Programme)
Global Environment Research Centre,
Imperial College, London

CLARENDON PRESS · OXFORD
1992

Oxford University Press, Walton Street, Oxford OX2 6DP
Oxford New York Toronto
Delhi Bombay Calcutta Madras Karachi
Petaling Jaya Singapore Hong Kong Tokyo
Nairobi Dar es Salaam Cape Town
Melbourne Auckland
and associated companies in
Berlin Ibadan

Oxford is a trade mark of Oxford University Press

Published in the United States
by Oxford University Press, New York

A catalogue record for this book is available from the British Library

Library of Congress Cataloging in Publication Data
Mason, B.J. (Basil John)
Acid rain: its causes and its effects on inland waters / B.J. Mason.
(Science, technology, and society series; 8)
Includes bibliographical references and index.
1. Acid pollution of rivers, lakes, etc. 2. Acid deposition — Environmental aspects — Europe. 3. Acid rain. I. Title. II. Series.
TD427.A27M37 1992 363.1'683 — dc20 92–3052

ISBN 0 19 8583443

Set by Graphicraft Typesetters Ltd, Hong Kong
Printed in Great Britain
by Bookcraft (Bath) Ltd
Midsomer Norton, Avon

Preface

In this small book I attempt to give a self-contained and integrated account of the results of some recent researches on the acidification of lakes and streams and the resulting toxic effects on fishes and other aquatic animals.

It is very largely based on the experimental and modelling studies of the recent five-year Anglo-Scandinavian Surface Waters Acidification Programme, described in Chapter 1.

In attempting to describe and explain the complex sequence of events and mechanisms that intervene between the emission of acidifying gases from a power station chimney to the toxic action of the acidified waters on the gills of a fish, I have relied heavily on the results of this strongly integrated and coordinated project, in which the key measurements were made simultaneously on a limited number of carefully chosen field sites by identical or directly comparable techniques.

In coordinating and interpreting the results of measurements made by many research groups working in different disciplines, one became conscious of the fact that the specialist literature is often confusing in regard to basic definitions, nomenclature, and terminology. I have therefore included such basic background information that may enable a non-specialist to follow the action, argument, and implications without having frequently to consult other texts.

I hope that I have also managed to convey something of what is involved in the planning and execution of complex field experiments that engaged some 300 scientists from about 30 institutions in three countries. For my part, I have greatly enjoyed the opportunity, after normal retirement, of working with, and learning from, colleagues in other disciplines. I much appreciated their forbearance in having a physicist in charge of what was predominantly a chemical and biological programme. That so much was accomplished in so short a period with modest resources is entirely due to their efforts. Any shortcomings or errors in this account of their findings are entirely my responsibility.

As described in Chapter 1, the whole programme derived from the foresight of Lord Marshall and was funded entirely by the former Central Electricity Generating Board and British Coal, with very few conditions and no interference.

It gives me much pleasure to record my warmest thanks for the unfailing help and support I have received from The Royal Society and the Norwegian and Swedish Academies, Sir Richard Southwood FRS and other members of the project's Management Group, and the Centre for Environmental Technology at Imperial College for providing me with a scientific base and office facilities during the entire project.

I am also greatly indebted to Mrs Jean Ludlam for producing an excellent typescript and helping with the index.

I am grateful to the Cambridge University Press for permission to reproduce many diagrams from *The surface waters acidification programme*, which presents the scientific results in full.

Thanks are due also to the Cambridge University Press, the Warren Spring Laboratory, the Meteorological Office, and other publishers and authors for permission to reproduce diagrams acknowledged in the text.

Imperial College, London
November 1991 B.J.M.

Contents

Plates

(Plates appear between pages 54–5)

1
The search for answers to key questions

1.1 Introduction

'Acid rain' is a general, short-hand term used to describe the deposition of all atmospheric pollutants of an acidic, or potentially acidic, nature whether they be deposited in rain or snow (wet deposition) or in the dry state as gases and small particles. But wet or dry, there is little doubt that acid deposition poses an ecological threat, especially to aquatic life in streams and lakes located on hard rocks and thin soils in southern Scandinavia and in some parts of Scotland and North America.

It may also contribute to the serious tree damage reported from Germany and several other European countries, but to what degree it is responsible is by no means clear and continues to be a matter of considerable controversy and debate. There appear to be several different symptoms of damage and several possible contributory causes with the overall impact varying from species to species. Climatic stress induced by severe droughts or frosts now appears to be an important factor, with acid rain playing a less dominant role than was once thought. However, firm, well-founded conclusions are not possible on the available evidence and are likely to come only from a more systematic and sustained research effort.

The situation is much clearer regarding the acidification of lakes and streams since publication of the results of the recent Anglo-Scandinavian Surface Waters Acidification Programme, which has achieved a remarkable degree of consensus in the three participating countries and has been influential in persuading the UK electricity-generating companies and the government to take remedial action.

1.2 The nature of the problem

This and similar problems become apparent when the damage exceeds a certain level generally accepted as normal and then spreads or intensifies as the natural control mechanisms fail to cope. It is then necessary to determine, by careful observation and measurement, the nature, extent, and intensity of the damage, the rates-of-change and whether these are gradual, episodic, or step-changes, and to compare these with past records if they exist. The next step is to correlate the damage symptoms with internal and external events judged to be likely causes or contributors. However, in most complex ecosystems there is unlikely to be a single well-defined cause but rather a combination of several contributory factors, some acting synergistically, others in contention, so that it becomes necessary to identify and study the controlling processes and mechanisms in what is usually a complex, interactive, multifactorial system. This may involve a combination of observational and experimental investigations in the field, laboratory and theoretical studies of basic processes, and simulation of the total system or parts thereof by means of mathematical models. Whatever the approach, progress will depend largely on ascertaining the facts from good, reliable observations.

1.3 The Surface Waters Acidification Programme

It was in this spirit that The Royal Society, the Norwegian Academy of Science and Letters, and the Royal Swedish Academy of Science, with funds provided equally by the Central Electricity Generating Board and British Coal, established a major five-year programme to study the acidification of streams and lakes in the three countries. The Surface Waters Acidification Programme (SWAP), with the present author as Director, got underway in 1984 and involved more than 30 research groups from a wide variety of disciplines and institutions working in a closely integrated and coordinated overall programme. Its objectives were to find answers to the following questions.

1. In the affected areas of Norway and Sweden, what are the factors, in addition to pH, that in practice determine the fishery status of lakes?
2. What are the biological, chemical, and hydrogeological characteristics of catchments which determine whether the composition of surface waters falls within a range acceptable to fish?
3. To what extent are these characteristics being adversely affected by the acid deposition itself?
4. What changes would be brought about in water chemistry and fishery status by given levels of reduction of man-made sulphur deposition?

The origins, rationale, design and management of the programme are described by Mason (1990, pp. 1–8).

The main goal of the research programme was to improve our knowledge and understanding of the physical, chemical, and biological processes involved in the acidification of streams and lakes and thereby make better predictions of future trends. The general strategy was to carry out detailed and extensive investigations on a limited number of sites in the three countries that would allow comparisons between highly acidified, pristine, and transitional catchments and between forested and unforested sites. It was already known, and SWAP has confirmed, that a broad correlation exists between the distribution and loading of acid deposition and the occurrence of acid lakes and streams. Moreover, strongly acidified surface waters with improverished fish populations tend to occur in areas with hard granite rocks and thin soils having a low cation-exchange capacity, whereas comparable levels of acid deposition cause little acidification if the soil is deep and rich in calcium. There was also evidence to suggest that acidic deposition on a catchment can be appreciably influenced by trees and other vegetation, whilst the chemistry of the run-off into lakes and streams may be affected by changes in land use.

The SWAP would therefore involve studies of catchment hydrogeology and vegetation, the chemical composition of the soils and underlying rocks, changes in the chemistry of the rain and snow water as it percolates through the different soil layers, and measurements of the water flow through the soil to determine the residence times available for the chemical and biological processes to act. Changes in the chemistry of stream waters would be studied in relation to the stream flow, with special attention given to the high flows that follow heavy rain and snow melt and are accompanied by pulses of high acidity and aluminium that are particularly toxic to fish. The last link in the chain would concern the effects of changes in water chemistry on fish at the various stages of their life history, and on other aquatic animals in the food chain.

A comprehensive and strongly integrated field programme, involving several different institutions and research groups, and structured as shown in Fig. 1.1, was established at highly acidified and transitional sites in all three countries and at a pristine site in northern Norway — see Figs. 1.2 and 1.3.

The field experiments were supplemented by laboratory studies of key processes such as the chemical weathering of rocks and soils, the speciation and toxicity of aluminium, the role of organic acids in surface-water chemistry, and the physiological reaction of fishes to water quality.

In view of the lack of a long time-series of direct measurements, it was decided to mount a major investigation into the past history of lake acidification in the three countries by analysing the distribution of acid-sensitive species of diatoms from radioactively dated lake sediments from which it is possible to reconstruct the pH profile of the sediments over the last few centuries. The detailed results of this work, involving some 20 sites and 15 research groups, were published in *Palaeolimnology and lake acidification* (Battarbee *et al.* 1990).

Several groups in SWAP exploited and further developed a Norwegian

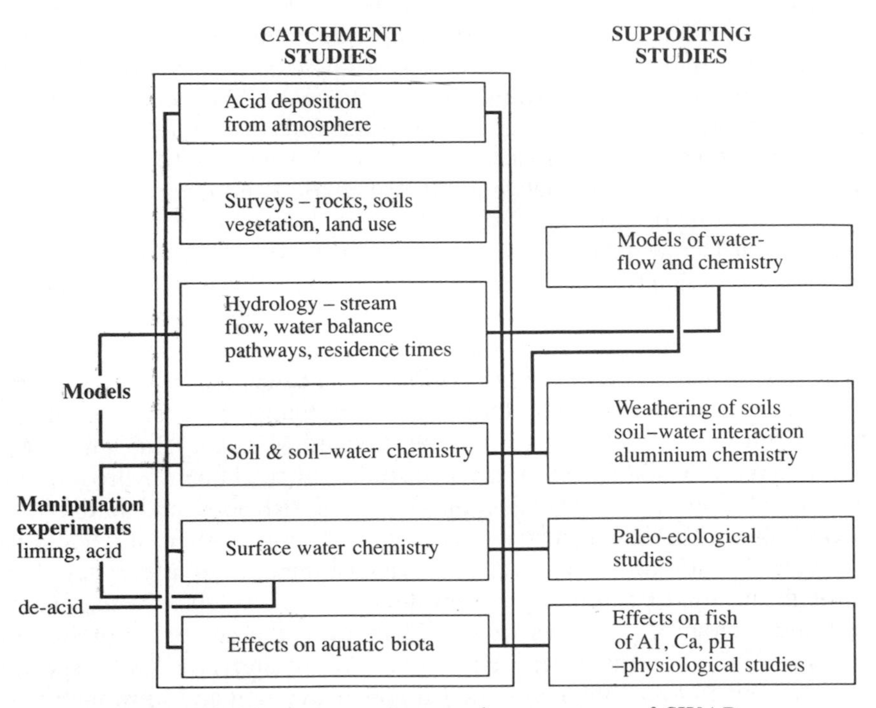

Fig. 1.1 The integrated research programme of SWAP.

mathematical model to simulate short-period changes in stream flow and chemistry, and to account for observed episodic and seasonal variations. Another model, developed in the USA, was adapted to stimulate long-term changes in surface water chemistry at a number of the experimental sites, and to infer what changes may occur in the future in response to various assumptions about future levels of acid deposition.

With so many institutions and research groups involved in the SWAP, it was necessary to take special measures to ensure compatability and direct comparability between the measurements, especially the chemical analyses, made by the different groups. Special workshops were therefore formed to develop standardized procedures, arrange for intercomparisons of methods, intercalibration of instruments, uniform methods of data analysis and quality control, to be adopted by all participants.

Workshops were also held to plan and coordinate studies of chemical weathering of rocks and soils, to standardize measurements of the various aluminium species, and to coordinate and discuss the various modelling studies.

A full description of all the research activities, their results and conclusions, appears in a substantial volume entitled *The surface waters acidification programme* (Mason 1990).

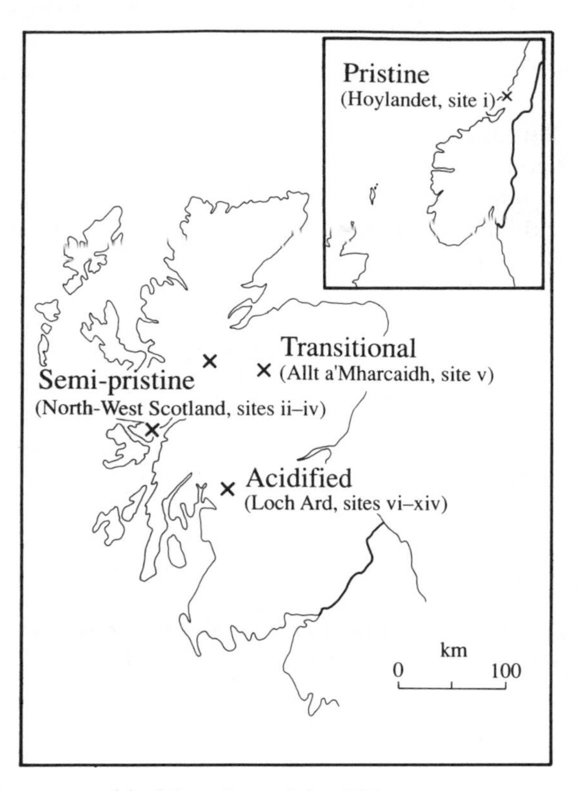

Fig. 1.2 The geographical location of the SWAP research sites in Scotland.

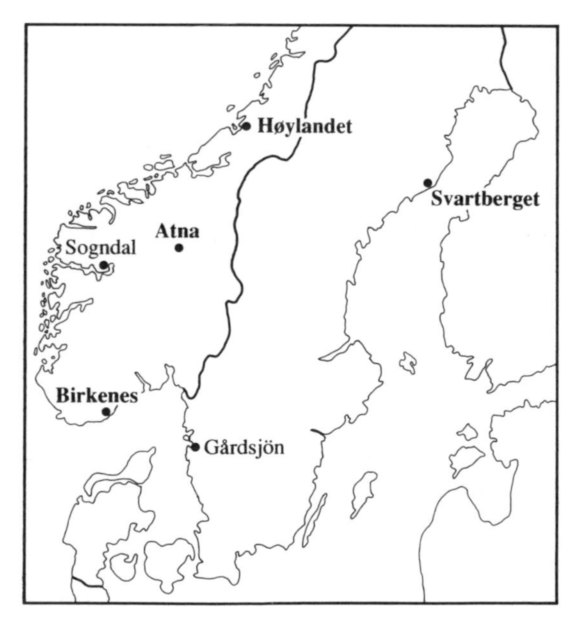

Fig. 1.3 The geographical location of the SWAP research sites in Scandinavia.

In this smaller monograph, the author attempts to synthesize the results of the SWAP and closely related research, and to present a self-contained account of the whole chain of events and processes that starts with the emission of acidifying gases into the atmosphere and ends with their toxic effects on fishes and other aquatic life.

1.4 References

Battarbee, R.W., Mason, B.J., Renberg, I., and Talling, J.F. (eds.) (1990). *Palaeolimnology and lake acidification*. The Royal Society, London.

Mason, B.J. (ed.) (1990). *The surface waters acidification programme*. Cambridge University Press.

FURTHER READING

The Royal Society (1990). Surface Waters Acidification Programme, Management group final report. *Science and Public Affairs,* **5**, 74–95.

2

The emission, transport, and chemical transformation of acid pollutants in the atmosphere

2.1 Introduction

Understanding the problem of acid deposition requires knowledge of the distribution in space and time of the major acidifying pollutants SO_2, NO_x, and HCl, their chemical transformation in the atmosphere, and their removal by deposition on the Earth's surface, either directly in gaseous or particulate form (dry deposition), or after incorporation into cloud and raindrops (wet deposition). The chemical reactions involved in both the gaseous and liquid phases are complex and incompletely understood. They are the subject of much active research involving the measurement of concentrations and conversion rates of chemical species in the atmosphere, laboratory measurements of key reaction rates, and the use of models to simulate the transport and deposition of the acidic substances between sources and receptor areas.

2.2 Emissions

Recent estimates of annual global natural sulphur emissions lie in the range 50–100 Mt S (megatonnes of sulphur) compared with man-made emissions of 60–80 Mt S. These estimates involve considerable uncertainty as they rely on extrapolation to a global scale from a relatively small number of measurements. Natural emissions arise in roughly equal proportions from terrestrial and marine sources, whereas anthropogenic emissions arise primarily from the combustion of fossil fuels in industrial processes. Of the latter, about 35 Mt S (around 40 per cent) originates from Europe, 19 Mt (24 per cent) from North America, and about 18 Mt (23 per cent) from Asia. Globally, SO_2 emissions from the combustion of fossil fuels continue to increase, the trend being most evident in industrially developing countries, whereas in western Europe emissions have decreased substantially since 1980.

Maps showing the 1988 emissions of SO_2, NO_x, and NH_3 for the whole of Europe, on a 150 km × 150 km grid, prepared by the European

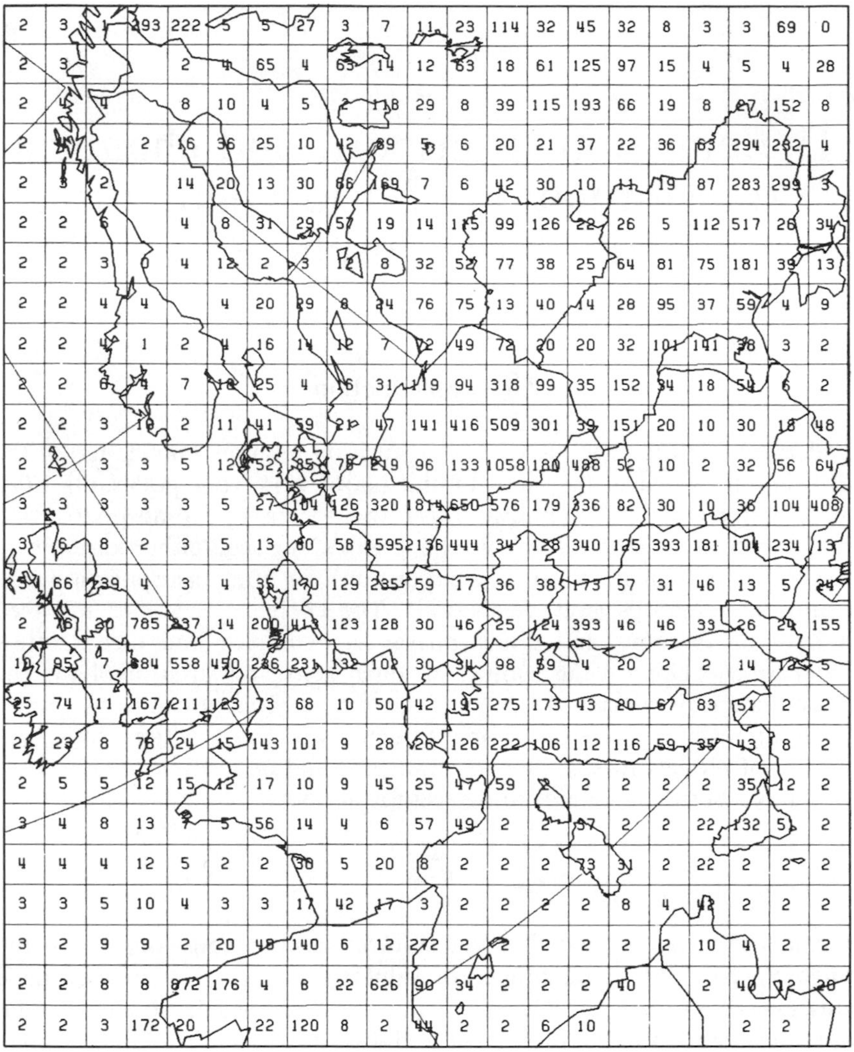

Fig. 2.1 1988 emissions of SO_2, in kilotonnes of sulphur per year (EMEP MSC-W 1990).

Monitoring and Evaluation Programme (EMEP), are shown in Figs. 2.1, 2.2, and 2.3. The largest values tend to lie in a belt extending from central England across Germany into southern Poland and Czechoslovakia, with other pockets in western Russia, northern Spain, and northern Italy. National emissions of SO_2, NO_x, NH_3, and VOC (volatile organic compounds, important in the generation of ozone) are given in Table 2.1 in kilotonnes per year (kt yr^{-1}).

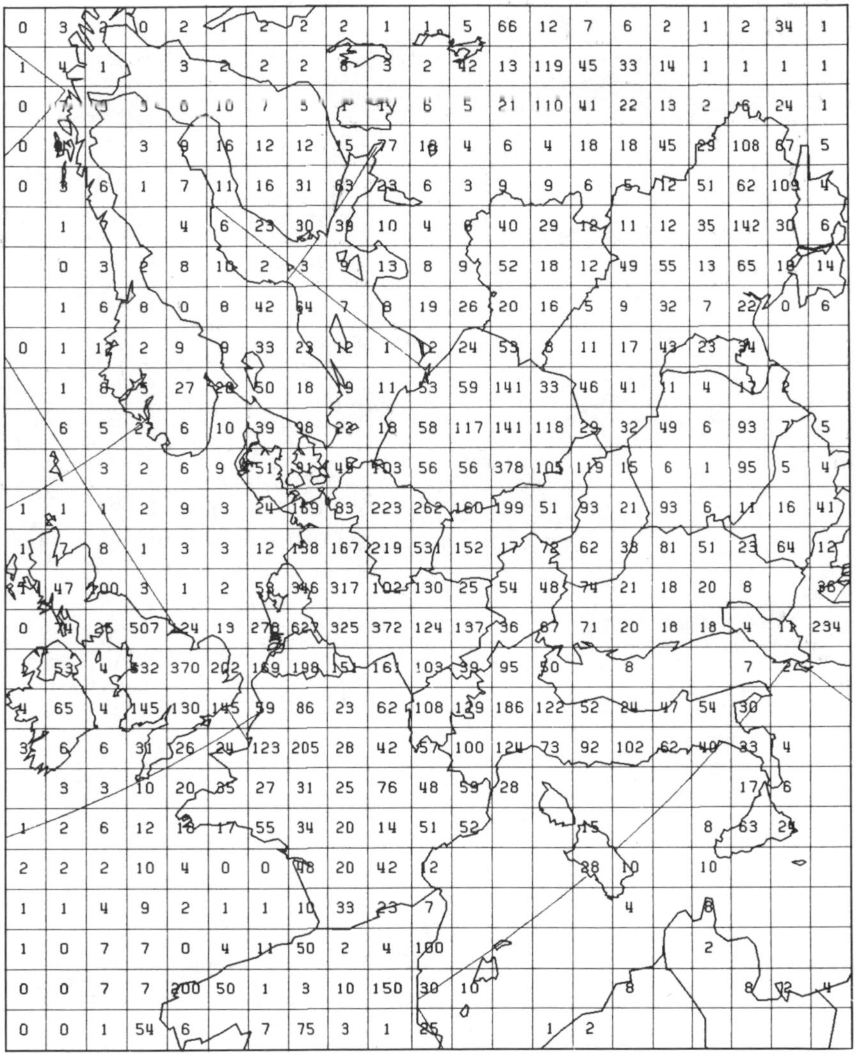

Fig. 2.2 1988 emissions of NO_2, in kilotonnes of sulphur per year (EMEP MSC-W 1990).

Sulphur dioxide. The trends in the emissions of SO_2 from the UK during this century are shown in Table 2.2 and, since 1970, in Fig. 2.4. The spatial distribution of emissions of SO_2 from the UK in 1983, based on 20 km × 20 km squares, is shown in Plate 1. The UK contributes less than 2 per cent to the total input of sulphur into the global atmosphere and about 8 per cent of the man-made sulphur emitted in Europe. UK emissions of SO_2, 75 per cent of which comes from power stations and

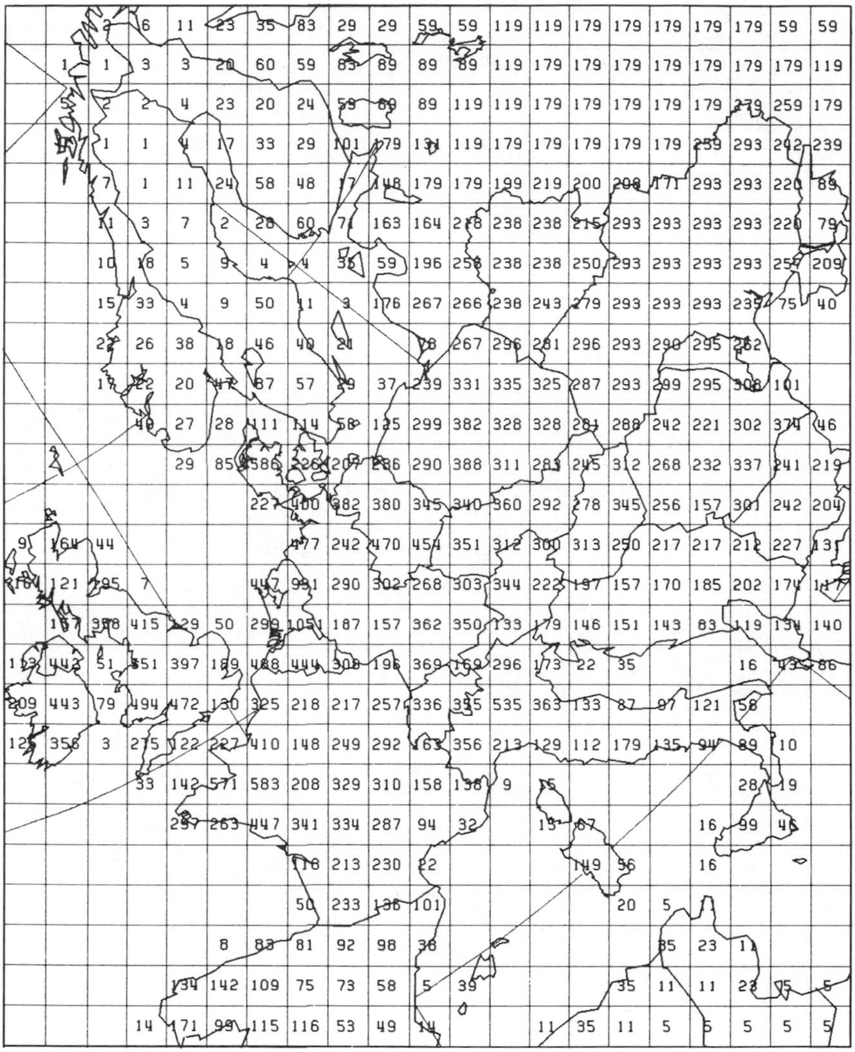

Fig. 2.3 1988 emissions of NH_3 in hundreds of tonnes per year (EMEP MSC-W, 1990).

about 20 per cent from other industrial plants, have fallen by 40 per cent since 1970 and by 25 per cent since 1980.

Nitrogen oxides. Total global emissions of NO_x are estimated to lie in the range 25–100 Mt N yr^{-1} (megatonnes of nitrogen per year) with natural sources accounting for perhaps one third. Emissions from the UK, about 30 per cent of which come from power stations and 45 per

Table 2.1 Total national emissions of SO_2 as sulphur, NO_x as nitrogen, NH_3 as nitrogen, and volatile organic compounds (VOC) in kilotonnes per year for 1985 (except for VOC data which refer to 1980).

Country	SO_2	NO_x	NH_3	VOC
Albania	25	3	20	—
Austria	85	66	70	391
Belgium	234	117	78	374
Bulgaria	570	46	121	—
Czechoslovakia	1575	373	165	—
Denmark	163	72	109	197
Finland	185	73	42	685
France	923	515	693	2971
Germany (East)	2500	291	200	—
Germany (West)	1200	882	361	2724
Greece	180	46	93	260
Hungary	710	91	125	—
Iceland	3	4	2	—
Ireland	69	21	114	94
Italy	1252	485	351	2116
Luxembourg	7	7	5	17
The Netherlands	138	163	140	534
Norway	50	68	34	319
Poland	2150	457	394	—
Portugal	134	58	45	436
Romania	100	119	288	—
Spain	1603	289	224	1786
Sweden	135	92	51	1201
Switzerland	48	65	51	361
Turkey	161	53	575	—
USSR	5550	892	2619	—
UK	1780	560	394	1746
Yugoslavia	588	58	194	—

From Smith 1991.

Table 2.2 Total emissions from the UK of SO_2 as sulphur and NO_x as nitrogen in megatonnes per year

	1900	1950	1960	1970	1980	1985	1989
SO_2	1.4	2.3	2.8	3.10	2.43	1.84	1.83
NO_x			0.4	0.73	0.74	0.70	0,76

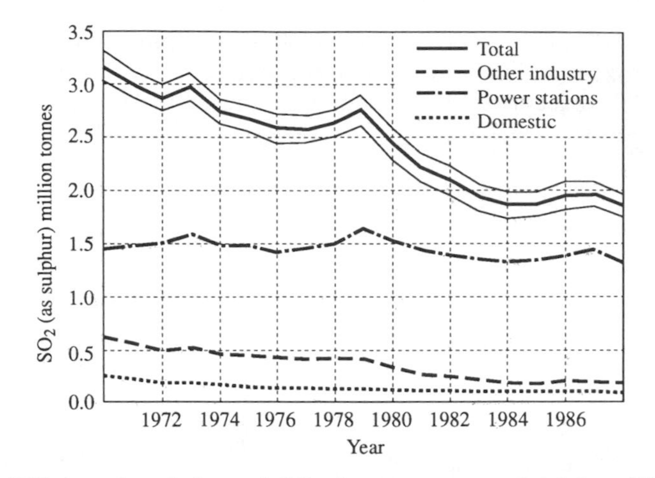

Fig. 2.4 UK annual emissions of SO_2, in megatonnes of sulphur. The lines on either side of the total curve represent 95 per cent confidence limits. (From Department of the Environment 1990, by courtesy of the Warren Spring Laboratory.)

cent from motor vehicles, account for about 10 per cent of the European total and are increasing only slowly, as shown in Fig. 2.5. The increase in vehicular emissions in recent years has been largely offset by a fall in the industrial uses of fuel oil. Natural emissions of NO_x from the soil are probably less than 10 per cent of anthropogenic emissions.

Hydrogen chloride (HCl). This is a highly soluble and reactive gas and, though of minor importance on a global or regional scale, it can significantly influence acid deposition close to sources. In western Europe, coal combustion is the major source, accounting for about 65 per cent of total emissions, with waste incineration being perhaps the next largest contributor. In the UK, coal combustion is estimated to contribute 93 per cent to emissions of HCl and waste incineration 6 per cent. The spatial distribution will depend not only on where and how much coal is burnt, but also on its chloride content, which can vary from less than 0.1 per cent to more than 0.4 per cent. The annual emissions from coal are 250–350 kt. In the absence of a detailed emissions inventory, it is not possible to make reasonably accurate estimates of HCl deposition.

Ammonia (NH_3). This is the major acid-neutralizing compound in the atmosphere and plays an important role in the chemistry of acid rain. The major sources are animal wastes and the application of fertilizers. In the UK, animal wastes are estimated to contribute about 80 per cent of the total emissions of 450 kt N yr^{-1}, cattle being responsible for as much

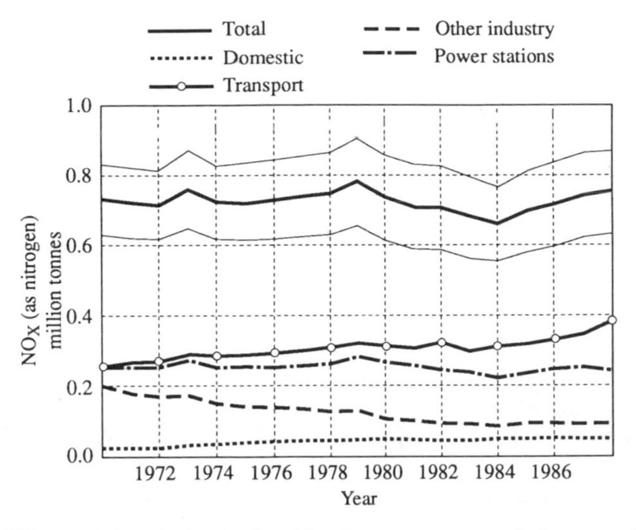

Fig. 2.5 UK annual emissions of oxides in megatonnes of nitrogen. The lines on either side of the total curve represent 95 per cent confidence limits. (From Department of the Environment 1990, by courtesy of the Warren Spring Laboratory.)

as 60 per cent. Ammonia emissions have increased in recent years with more intensive animal husbandry. For example, over Europe, NH_3 emissions from livestock increased by 50 per cent between 1950 and 1980, the largest increases occurring in The Netherlands, where emissions have more than doubled.

2.3 Transport and chemical transformation of acidifying gases

Once emitted into the atmosphere, the pollutants are carried, dispersed, and diluted by atmospheric motions having a wide range of scales. The plume from a point source, such as a power station chimney, spreads out into an expanding cone which meanders with fluctuations in the wind. For the most part, the plume is confined within the well-mixed atmospheric boundary layer, the depth of which may vary from a few hundred metres to about 2 km, depending on the time of day and the prevailing weather system. Ultimately, the plume will lose its coherence and shape as it becomes disrupted and distorted by convective motions and wind shear within the boundary layer. Within the mixing layer, emission plumes are dispersed both horizontally and vertically so that pollution concentrations decrease steadily with distance downwind. Thus, by the time a plume from a chimney 200 m high first contacts the ground some 10–20 km downwind, the concentration will be typically 10 000 times less than in the flue gas.

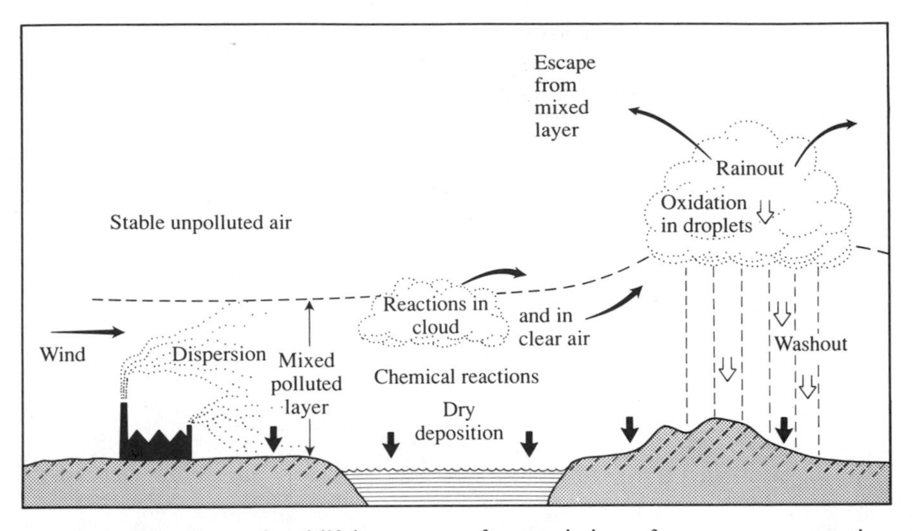

Fig. 2.6 The fate of acidifying gases after emissions from a power station chimney. (With modifications from CEGB 1987).

Beyond the point of contact with the ground, a good deal of the acidic pollutant is deposited on the surface or vegetation in dry (gaseous or particulate) form by absorption, impaction or sedimentation. The rest may travel for hundreds or even thousands of kilometres. During this travel time of up to a few days, the gases SO_2 and NO_x are oxidized and converted into sulphuric and nitric acid, either in gas-phase reactions or, more effectively, by being captured by cloud and raindrops (where the chemical transformations proceed much more rapidly in the liquid phase), and are eventually brought to the surface in rain or snow. These features are summarized in Fig. 2.6.

The rate of dry acidic deposition is usually estimated by the product of the ground level concentration of the precursor gas (SO_2 or NO_x) and the 'deposition velocity'. Measurements have shown that the deposition velocities vary from one surface to another (e.g. from bare soil to short grass to crops, etc.), and also depend on the stability of the air close to the ground, and therefore differ with time of day, weather, and season. For a typical deposition velocity of $0.8\ \mathrm{cm\,s^{-1}}$, the amount of SO_2 uniformly mixed in a boundary layer 1 km deep will be depleted by about 3 per cent per hour. The actual flux or intensity of the dry acid deposition will decrease with increasing distance from the source, and how much remains to be deposited in this form depends on the extent to which the plume encounters cloud and rain and the rate at which the SO_2 is oxidized to sulphate during its travel.

The rates of removal of sulphur from the atmosphere during rain are generally much greater than the rates of dry deposition. Moderate rain will often remove more material in one hour than will dry deposition over 2–3 days. However, the sulphur content of rainwater is far greater than can be accounted for by the dissolution of SO_2, which is limited by the fact that this increases the acidity of the water to a point where no more can dissolve. This self-limiting process allows only about 1 per cent of atmospheric SO_2 to be removed in this way, and points to the necessity of the SO_2 being oxidized to sulphate before deposition.

Although acid production proceeds much more rapidly in the aqueous phase, clouds and rain are present only a few per cent of the time, so gaseous transformations and deposition are important and account for about two thirds of the total acid deposition in the UK, and about one third in southern Norway.

The rates of conversions of SO_2 and NO_x to H_2SO_4 and HNO_3 have been determined by measurements from the Hercules (C-130) flying laboratory of the UK Meteorological Office and by a smaller aircraft operated by the Central Electricity Generating Board. The aircraft can locate and follow a chemically marked plume from a particular power station, sample the air inside and outside the plume as it spreads downwind, and analyse it for all the relevant chemical species (e.g. SO_2, NO, NO_2, oxidizing agents such as O_3, H_2O_2), hydrocarbons, and aerosols. Cloud and rainwater samples are collected and analysed for pH, all the main ionic species, H_2O_2, and other chemicals. In order to explain the measured conversion rates of SO_2 and NO_x into acids, it is necessary to invoke photochemical reactions involving highly reactive oxidizing agents such as O_3, leading to the formation of the important radical OH, which is unreactive with oxygen and therefore relatively stable. Some of the more important chemical reactions can be summarized as follows.

Gaseous reactions in a dry atmosphere

Sulphuric acid

$$SO_2 + OH + M \rightarrow HSO_3 + M$$
$$HSO_3 + O_2 \rightarrow HO_2 + SO_3$$
$$SO_3 + H_2O \rightarrow H_2SO_4,$$

with HO resulting from the photolytic decomposition of ozone and reaction with water vapour, i.e.

$$O_3 + h\nu \rightarrow O^* + O_2$$
$$O^* + H_2O \rightarrow 2HO,$$

where M denotes a third non-reactive molecule, such as N_2, and * denotes the excited atom. The aircraft measurements indicate a conversion rate of about 4 per cent per hour on a clear summer's day, when solar ultraviolet radiation permits ready photolysis of ozone, but the rate is much reduced in winter. A typical 24-hour average value is 1 per cent per hour.

Nitric acid. NO is oxidized rapidly in sunlight by O_3 to form NO_2

$$NO + O_3 \rightarrow NO_2 + O_2,$$

and this reacts with the OH radical to form nitric acid vapour, i.e.

$$NO_2 + OH + M \rightarrow HNO_3 + M.$$

The conversion rate is about ten times faster than for the corresponding SO_2 reaction, so conversion would be virtually complete during a traverse of the plume across the North Sea in summer. At night, conversion may take place via the formation of the nitrate radical, which is photochemically unstable in daylight. The reactions are

$$NO_2 + O_3 \rightarrow NO_3 + O_2$$
$$NO_2 + NO_3 + M \rightarrow N_2O_5 + M$$
$$N_2O_5 + H_2O \rightarrow 2HNO_3;$$

the last step is likely to take place on the surfaces of aerosol particles carrying an adsorbed water film.

Liquid-phase reactions in clouds and rain
Sulphuric acid

$$2SO_2 + 2H_2O \rightarrow SO_3^- + HSO_3^- + 3H^+$$
$$HSO_3^- + H_2O_2 \rightarrow HSO_4^- + H_2O,$$

the highly soluble H_2O_2 resulting from

$$HO_2 + HO_2 \rightarrow H_2O_2 + O_2$$

and HO_2 by the photolysis of carbonyl compounds, for example

$$HCHO + h\nu \rightarrow H + HCO$$
$$HCO + O_2 \rightarrow HO_2 + CO.$$

The conversion rates of SO_2 to SO_4, which are independent of pH, are very fast, almost 100 per cent per hour in summer but, in many situations, may be limited by the supply of hydrogen peroxide which, in polluted air, is often present in small concentrations compared with those of SO_2. The conversion rates may therefore be highly non-linear with respect to SO_2 concentration. This suggests that a reduction in SO_2 emissions may not lead to a proportionate reduction in wet acid deposition.

Nitric acid. As both NO and NO_2 are only very slightly soluble in water, the nitric acid content of rain is thought to be due largely to the adsorption of nitric acid vapour and the capture by raindrops of aerosols containing N_2O_5. The latter results in the following reaction:

$$N_2O_5 + H_2O \text{ (liq)} \rightarrow 2HNO_3$$

which is believed to be rapid, but conversion rates have not yet been established.

Both the aircraft measurements and the photochemical models indicate that the rates of production of acids from the precursor gases are often limited by the availability of oxidizing agents, hydrocarbons, and solar ultraviolet radiation.

2.4 References

CEGB (1987). *Acid rain*, CEGB Research Bulletin No. 20.

EMEP (European Monitoring and Evaluation Programme) MSC-W 1990, Norwegian Meteorological Institute, Oslo.

Smith, F.B. (1991). An overview of the acid rain problem. *Meteorological Magazine*, **120**, 77–91.

Department of the Environment (1990). *Acid deposition in the United Kingdom, 1986–8*. Published by the Warren Spring Laboratory for the Department of the Environment, London.

FURTHER READING

Wayne, R.P. (1991). *Chemistry of atmospheres*, (2nd edn), Chapter 5. Oxford University Press.

3
Acid deposition in the UK and Europe

3.1 Introduction

In order to detect and assess the effects of changes in the emissions of SO_2, NO_x, and their oxidants on the acidity of precipitation and on acid deposition, it is necessary to have long-term, reliable, representative, and accurate measurements of pH, alkalinity, and concentrations of the main ionic species. Very few measurements made before 1980 satisfied these criteria, but since then considerable efforts have been made, notably in the UK and Scandinavia, both to increase the number of observing sites and to establish collaborative arrangements for improvements in techniques, calibration, and intercomparison using standardized methods and protocols in order to obtain good precision and reproducibility of measurements under field conditions. This was a difficult but essential task.

The pH of rainwater is a measure of its acidity and is defined as the negative logarithm of the molar hydrogen ion concentration:

$$pH = -\log_{10}[H^+],$$

so that a change of one pH unit represents a tenfold change in H^+ ion concentration. The pH of pure water, in which H^+ and OH^- ions are produced only by the thermal dissociation of water molecules, is 7.0, or $[H^+] = 10^{-7}$ M. The pH of 'pure water' in equilibrium with atmospheric carbon dioxide is 5.6. Values lower than this for rainwater are found even in remote parts of the world far from man-made sources of pollution. This suggests either world-wide dispersion of such pollution or the presence of natural sources of sulphur that are readily oxidized to produce a natural background of sulphate. Measurements of the pH of rain and melted snow in remote regions suggest a base-line value of pH 5.0 for precipitation in the absence of anthropogenic sources of acidifying agents.

3.2 Acid deposition in the UK

A network of some 60 monitoring sites has recently been established in the UK to measure the chemical composition of precipitation and the concentrations of acidic species in the air near the ground from which may be derived values of both wet and dry acid deposition. A comprehensive report on these measurements and computations, with maps showing the spatial distribution of the various parameters, has recently been published by the Department of the Environment (1990).

3.2.1 WET DEPOSITION

The patterns of wet deposition are based on weekly measurements using 60 open gauges which collect both wet and dry deposition. In addition, nine collectors, which open only when it rains, provide daily measurements at a number of representative sites; these have been used to define the ranges of concentrations of the various chemical species in precipitation occurring in different parts of the country.

Wet deposition is less easy to map than concentrations in precipitation because it exhibits greater variability. For example, areas of large wet deposition occur both in areas of comparatively low rainfall close to major sources and also in remote areas with high rainfall. There are comparatively small areas where the long-term concentrations of acidity, sulphate, and nitrate exceed five times the average. Acidity of rainfall in remote forested areas may be considerably enhanced by organic acids or may be reduced in areas subject to windborne alkaline dusts. Accordingly, maps of acid deposition require careful interpretation in terms of weather patterns, rainfall, wind fields, sunshine, and topography, taking into account any special features of both source and receptor areas.

Interpretation is also complicated by the fact that wet deposition is generally highly episodic in character with a high proportion of the annual total occurring on only a few wet days. If we define episodicity as the percentage of wet days that accounts for 30 per cent of the annual deposition, then this varies across the UK from about 3 per cent to 8 per cent, depending on the location of the site in relation to major source areas and the main directions of the rain-bearing clouds.

Precipitation-weighted mean values for the concentration of non-marine sulphate for 1986–8 are shown in Plate 2. The largest concentrations of 80–110 μeq l^{-1} (microequivalents per litre)* occur in the east Midlands and south Yorkshire and the lowest values of 10–20 μeq l^{-1} are found in north-west Scotland. There is a marked east–west gradient across the country. Over most of the country these values, averaged over 20 km $\times$ 20 km squares, are estimated to be reliable within $\pm$30 per cent. As most precipitation occurs in westerly or south-westerly airflows, elevated sulphate concentrations are to be expected to the east and north-east of the major source areas, but this simple picture is complicated by rather infrequent but large influxes from Europe.

* The concentration of an ionic solution in μeq l^{-1} is related to that expressed in μg l^{-1} by 1 μg l^{-1} $\equiv$ z/W μeq l^{-1}, where z is the ionic charge (1, 2, 3, etc) and W the molecular weight of the element.

The mean annual wet deposition of non-marine, mostly anthropogenic, sulphate, obtained by multiplying the interpolated values of concentration by the measured rainfall, is shown in Plate 3. The areas of greatest deposition are in the east Midlands and the Pennines close to major sources, and in the mountainous western areas of Wales and central Scotland where the concentrations are low but the rainfall is high. These exceed 1.2 g S m^{-2} whereas the areas of lowest deposition in south-west England and north-east Scotland receive less than 0.6 g S m^{-2}.

The distribution of *total* wet sulphate deposition is shown in Plate 4. which includes the marine sulphate and elevates the values on average by about 0.4 g S m^{-2}.

Maps of mean nitrate concentration and annual wet nitrate deposition are shown in Plate 5. The distribution patterns are generally similar to those of non-marine sulphate, the nitrate concentrations being about half those for sulphate. The nitrate map does not show an elevated concentration in the East Midlands because the power stations contribute a much smaller proportion of the national output of NO_x than is the case for SO_2. The highest depositions of total sulphate and nitrate are among the largest in Europe.

Mean concentrations and annual mean depositions of the H^+ ion are shown in Plate 6. The distribution of acidity reflects a balance between the concentrations of acidic anions, mainly SO_4^{2-} and NO_3^-, and of basic cations such as NH_4^+ and Ca^{2+}. The most acidic rainfall (greater than 60 μeq l^{-1}, pH less than 4.2) occurs in the east Midlands and Yorkshire. In contrast, the acidity in the northern and western coastal areas is close to the global background value of 10 μeq l^{-1}, pH 5.0. The deposition of acidity again reflects both the distribution of sources and rainfall amounts enhanced by orography. The fact that deposition in southern England is small relative to that in eastern Scotland although the depositions of sulphate and nitrate are similar, is explained by the larger concentrations of NH_4^+ and Ca^{2+} in southern England.

Comparison of the measurements made in 1981–5 and those made in 1978–80 shows that, while the geographical distribution of acidity remained much the same, the average H^+ ion concentration of the rain decreased by about 25 per cent, and that of the non-marine sulphate by rather less, between the two periods. During the same interval, the UK emissions of SO_2 decreased by 25 per cent. Measurements made at a considerably larger network of stations during 1986–8 showed no further significant decrease in H^+ ion concentration, consistent with the fact that SO_2 emissions had not changed appreciably in the meantime.

The wet deposition of non-marine sulphate, nitrate, ammonium, and acidity is highly episodic, only 3–8 per cent of wet days contributing 30 per cent of the total. There is also a large seasonal variation in the concentrations of the major ionic species of anthropogenic origin. Whilst concentrations are largest in late spring and smallest in early winter, rainfall amounts show a nearly inverse variation so that the overall effect on deposition in areas of high rainfall is small. Enhanced photochemical oxidation or increases in emissions of marine biogenic sulphur may contribute to the higher concentrations of sulphate in the late spring, but seasonal variations in windflow and the arrival of rain-bearing clouds also contribute to the variability.

The above estimates of wet deposition are made on the assumption that the concentrations of major ion species in precipitation remain constant with altitude. However, in mountainous terrain, a substantial fraction of the rainfall results from the collection of cloud water from orographic cloud caps by raindrops falling in from higher levels. Since the stationary cap clouds are continually fed with low-level air, the cloud water may have consistently higher concentrations of pollutants and so enhance the concentrations in rainwater arriving on the mountain slopes. Measurements made at Great Dun Fell in the northern Pennines show that the rainfall roughly doubled in the height interval 200 m to 850 m above sea level and that the wet deposition increased by between a factor of 4 and 6. It has been estimated that in the high rainfall areas of Snowdonia, Cumbria, Galloway, and the west-central Highlands, the wet deposition is increased by some 70 per cent owing to these effects. Because high mountains cover only a rather small area of the UK, this enhancement may not significantly affect national budgets but it may have a considerable impact locally, especially in the most acid-sensitive fresh-water catchments.

The EMEP model, described in the next section, has been used to compute air trajectories arriving over the UK at 6-hourly intervals and hence to compute the wet deposition accompanying these trajectories at one site, Stoke Ferry in Norfolk. The contributions of the various source regions to this deposition were then calculated and attributed, as shown in Table 3.1, where the undefined category included trajectories which passed over both UK and European sources, and for which separate contributions could not be assigned. At this site at least 60 per cent, and probably 80 per cent, of the deposition of anthropogenic sulphur originated in the UK.

3.2.2 DRY DEPOSITION

Dry deposition of SO_2, as calculated from the product of the concentration in the air near the ground and the deposition velocity, computed as a function of the nature and roughness of the surface, is plotted in Plate 7. The largest values (more than 3 g S m^{-2} yr^{-1}) occur in the industrial midlands of England. In general, the pattern closely follows that of the SO_2 concentrations in the air. Dry deposition of NO_2 is slow and is not a major removal mechanism for NO_x, except in south-east England. Nitric acid gas has a high deposition velocity but concentrations are low so this does not contribute substantially to the total nitrogen deposition.

Table 3.1 Percentage contributions of source regions to wet deposition at Stoke Ferry

	SO_4^{2-}*[a]	NO_3^-	NH_4^+	Rainfall
UK	70	66	69	76
Europe	9	10	8	6
North	4	3	4	4
Undefined	18	20	20	14

[a] Non-marine.

3.2.3 TOTAL DEPOSITION

The total annual deposition of *sulphate*, computed for 20 km × 20 km squares, is shown in Plate 8. These values represent the sum of wet deposited sulphate (including marine sulphate and enhancement by orographic clouds), dry deposited SO_2, and estimates of sulphate deposited in cloud and fog water. The deposition of aerosol sulphate, considered to be small, is neglected.

The annual deposition ranges from more than 6 g S m^{-2} in the southern Pennines to less than 1 g S m^{-2} in parts of northern Scotland. Over parts of central and southern England, dry deposition is the major component, but over most of the country, wet deposition dominates and accounts for more than 80 per cent of the total in many western and northern areas. Although there may be pockets of greater uncertainty, on average the 20 km square estimates are believed to be accurate to within a factor of 2.

Estimates of the total deposition of oxidized nitrogen species are mapped in Plate 9. These include wet deposition of nitrate with orographic cloud enhancement, dry deposition of NO_2 and HNO_3, and the deposition of nitrate in cloud and fog water. The total annual deposition varies from more than 0.4 g N m^{-2} in northern Scotland to 1.2 g N m^{-2} in the wetter regions of northern England and Wales. In south-east England, dry deposition accounts for about half of the total, but elsewhere wet deposition dominates and in mountainous areas may account for 90 per cent of the total.

In considering total nitrogen deposition, it is necessary to take into account that over much of the country dry deposition of ammonia is believed to dominate. Unfortunately, large uncertainties exist in the concentrations of ammonia and in our understanding of its deposition. Over forested areas, this is estimated to vary from about 3 g N m^{-2} in northern Scotland to as much as 7–8 g N m^{-2} in southern England. Over moorland, depositions are much less, ranging from about 1 g N m^{-2} in northern Scotland to around 4 g N m^{-2} in Wales and Cumbria.

Table 3.2 Total nitrogen deposition at Eskdalemuir in grams of nitrogen per square metre per year

	Forest	Moorland
Wet deposition		
NO_3^-	0.33	0.33
NH_4^+	0.35	0.35
Dry deposition		
NO_2	0.05	0.05
HNO_3	0.2	0.06
NH_3	2.4	0.8
Cloudwater deposition		
NO_3^-	0.06	0.02
NH_4^+	0.06	0.02
Total	3.4	1.6

The results of a special study to determine the components of the total nitrogen deposition at one site, Eskdalemuir on the Scottish border, are shown in Table 3.2. Here wet deposition to moorland accounts for about 80 per cent of the total sulphur but only about 42 per cent of the total nitrogen deposition. The corresponding figures for the forested areas are 77 and 20 per cent. The much greater inputs of NH_3 and HNO_3 in dry deposition over the forest suggest that nitrogen inputs may double as the result of afforestation and account for up to two thirds of the total. However, these results are likely to be site specific and should not be extrapolated to other areas.

3.2.4 THE EPISODIC CHARACTER OF DEPOSITIONS

The daily average concentrations or depositions of pollutants at any monitoring site show large fluctuations in magnitude. For example, in central England, 30 per cent of the wet deposition of sulphate falls on only about 5 per cent of wet days. In Europe this latter figure, termed the episodicity, varies from about 3 to 15 per cent. The episodicity depends on the location of the monitoring site in relation to the major source areas and the main directions of the rain-bearing winds. Episodes of either high deposition or concentrations in rainwater are often associated with particular weather patterns. For example, pollution may build up within the air of a stagnant anticyclone and then be drawn into the circulation of a mobile depression where it is washed out by the rain.

Based on an analysis of daily measurements of the concentrations of non-marine sulphate in precipitation over Europe, Fig. 3.1 shows the

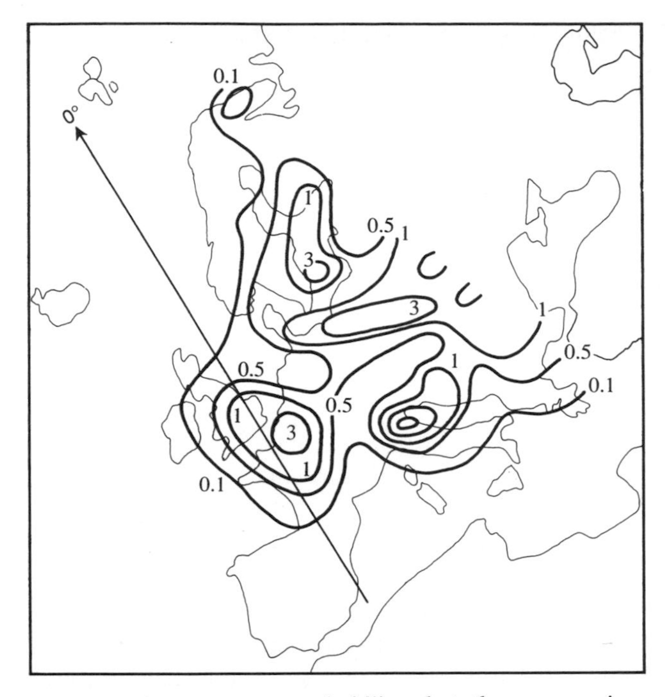

Fig. 3.1 Contours of percentage probability that the non-marine sulphate in precipitation will exceed 10 mg S l^{-1}. (From Smith 1991, by permission of The Royal Meteorological Society.)

contours of the percentage probability that these concentrations will exceed 10 mg S l^{-1}. The highest probabilities occur in Italy, Poland, and northern France, and also in southern Sweden, where they are likely to be damaging in this highly sensitive area. Fortunately, these episodes of high concentration are much less frequent in Norway and Scotland.

3.3 Modelling and measurement of acid deposition over Europe

The simple calculations quoted in Section 2.3 for the rate of deposition of sulphur from plumes confined to the atmospheric boundary layer suggest that much of the sulphur emitted by one country may be deposited in another. In order to quantify such international exchanges on a long-term basis, it is necessary to develop mathematical models of this long-range transport of pollutants.

Given the location and the strength of the sources of SO_2 (and NO_x), and having determined the air trajectories, say every 6 hours, from operational weather-forecasting models, the problem is to compute the changes in the concentrations of sulphur species in the air parcels that follow these trajectories. In practice, high precision in the location of individual trajectories is not required in order to calculate annual average concentrations or depositions. Having made these computations it is possible to draw up import–export budgets for each country allowing, in principle, the total deposition in any one country to be attributed percentage-wise to the countries of origin. A cooperative European programme on long-range transport of airborne pollution (EMEP) was established for this purpose in 1977 — see Eliassen *et al.* (1988) and EMEP (1987).

In effect, the transport model, of which there are several versions, calculates the changes in the concentrations of SO_2 and particulate sulphur in air parcels that follow air trajectories in the atmospheric boundary layer and eventually terminate in a particular 150 km × 150 km grid square on a map of Europe, in which both the dry and wet deposition of sulphur are calculated. The varying depth of the boundary layer is determined from meteorological soundings of temperature- and wind-gradients and allowance is made for the leakage of material from the top of the boundary layer into the free atmosphere. The physical and chemical loss and transformation processes are rather crudely represented by constant, uniform parameters derived from measurements of sulphate in rainwater and of SO_2 levels in the atmosphere on the assumption that there is a simple relationship between them.

Given the sources of SO_2 emissions listed in Table 2.1, the model computes the country-by-country sulphur budgets and depositions. The major contributors to the depositions in Norway and Sweden are shown in Table 3.3. However, these attributions involve considerable uncertainties because the computed values of deposition, when compared with those deduced from measurements, reveal major discrepancies in that up to 30 per cent of the deposition in Sweden and up to 50 per cent of that in Norway cannot be accounted for by the computed imports from European countries. Moreover, the percentage imports from the various countries may vary a good deal from year to year, as is seen in Table 3.3, because of differences in weather patterns, especially in wind flow and precipitation. Some of the unidentified deposition is thought to originate from outside Europe, for example, from North America, or from leakage out of the top of the boundary layer, thus contributing to a general background of pollution.

It has also been suggested that some of the background concentration of sulphur may be attributed to the emission of dimethyl sulphide (DMS, $(CH_3)_2S$) from the spring and early summer blooms of phytoplankton in the oceans. In the atmosphere DMS is rapidly oxidized, principally by reactions with OH and HO_2 radicals to form methyl sulphonic acid, CH_3SO_3H. Model calculations have suggested that this could contribute

Table 3.3 Percentage contributions of emitter countries to non-marine sulphate deposition in Norway and Sweden

	Sweden		Norway	
	1986	1988[a]	1986	1988[a]
Denmark	4	4	2	3
DDR	15	18	7	13
Czechoslovakia	6	5	3	3
FRG	5	4	5	4
Great Britain	6	8	11	17
Poland	9	11	4	8
Sweden	12	12	1	2
USSR	5	9	5	9
Indeterminate	31	16	52	21

[a] 1988 values are preliminary estimates.

as much as 30 per cent of the total sulphur deposition over southern Scandinavia during the season of plankton blooms, but the unpredictability of the location and timing of the blooms makes it difficult to test this hypothesis.

Despite these uncertainties, it seems probable that the UK is the largest single contributor to sulphur deposition in Norway and is responsible for up to 20 per cent in some years. More than half the sulphur deposition in Sweden originates in eastern Europe, the UK being a significant but not a major contributor.

3.4 Critical loads and target levels for acid deposition

The need to reduce sustantially the emissions of sulphur and nitrogen oxides in Europe over the next decade having been accepted by many governments, there is considerable debate on how this might best be achieved. Various strategies have been devised based, for example, on the following: flat rate reductions; varying percentage reductions for different countries; larger percentage reductions (up to 60 per cent for SO_2 and 30 per cent for NO_x) for large power plants; implementation of the best available technology. In the search for a more targetted and therefore perhaps a more cost-effective approach, effort has concentrated recently on deriving target deposition levels based on the concept of critical loads for particular areas, depending upon their vulnerability or sensitivity to acid deposition, as described by Kuylenstierna and Chadwick (1989).

The first step involves assessment of the sensitivity of a receptor area and placing it in one of, say, five categories based on a weighted combination of such factors as bedrock geology (e.g. weathering rates), soil properties (e.g. pH and base saturation), land-use, and rainfall amount. A major difficulty is that a receptor area such as a 150 km × 150 km EMEP grid square is likely to be far from homogeneous in regard to topography, geology, soil structure and composition, drainage, and vegetation. There follows an even more difficult task of specifying critical loads for each receptor area defined, rather idealistically, by UN-ECE in 1988 as 'A quantitative estimate of an exposure to one or more pollutants below which significant harmful effects on specified sensitive elements of the environment do not occur according to present knowledge.' This assumes that there is a threshold level of deposition below which harmful effects are unlikely to occur or cannot be distinguished from natural variations.

Although field observations or experiments might provide reasonable estimates of critical loads for a very limited number of ecosystems or species, this is highly unlikely to be the case for a large area containing many systems and species of widely differing sensitivities. Consequently, the assignment of a single critical load may not protect small but important ecosystems or the most endangered species. On the other hand, working at much finer resolution, with much smaller receptor areas, may be impracticable because the long-range transport of pollutants allows control, monitoring, and measurement of deposition on only a coarse mesh, while much the same is true for the sampling of rocks, soils, rainfall, water flow, etc.

Furthermore, the whole concept of critical loads and differential sensitivity to acidic deposition lacks a firm theoretical base, as demonstrated by our present inability to make reliable predictions of how soils, surface waters, and ecosystems will respond to changes in acidic deposition, bearing in mind that a large fraction of the annual load and damage may occur on only a few days of heavy rain or snow melt.

Notwithstanding these difficulties, target levels have been proposed for Europe and parts of the USA ranging from 20 keq H^+ km^{-2} yr^{-1} for the most sensitive areas to more than 160 keq H^+ km^{-2} yr^{-1} for the least sensitive areas.

Whether such targets will lead to more acceptable and effective emission strategies than national quotas, or percentage reductions, remains to be seen, but they are likely to be difficult to implement quickly in Europe, where they call for sulphur depositions to be reduced to one-quarter in the most sensitive regions, such as southern Scandinavia, where one third to one half of the deposition is contributed by eastern European countries.

3.5 References

Department of the Environment (1990). *Acid deposition in the United Kingdom, 1986–8*. Department of the Environment, London.

Eliassen, A., Hov, O., Iversen, T., Saltbones, J., and Simpson, D. (1988). *EMEP report 1/88*. Norwegian Meteorological Institute, Oslo.

EMEP (European Monitoring and Evaluation Programme) (1987). *Sulphur budgets in Europe for 1979–85*. Norwegian Meteorological Institute, Oslo.

Kuylenstierna, J.C.I. and Chadwick, M.J. (1989). In *Regional acidification models*, (ed. J. Kamari), Chapter 1. Springer, Berlin.

Smith, F.B. (1991). *Quarterly Journal of the Royal Meteorological Society*, **117A**, 657–83.

FURTHER READING

Smith, F.B. (1991). *Meteorological Magazine*, **120**, 77–91.

4
The role of hydrology in determining the chemistry of surface waters

4.1 Mechanisms of water transport through soils

On entering the soil, rain or snow melt-water may be stored or move towards the stream channel, depending upon the structure and the antecedent water content of the soil. Water movement occurs in response to a hydraulic potential gradient generated by the combined effects of gravitational head, capillary pressure, and the osmotic pressure of the soil water. The hydraulic conductivity of a soil, which determines the speed with which the water front moves through the soil profile under a given hydraulic pressure gradient, depends upon the pore geometry and continuity, the soil texture, and water content. If the water content is low, only the smaller pores are likely to be filled and these transmit water much less readily than the larger pores. The incoming water changes the water content of the surface soil until the hydraulic flow matches the rainfall intensity and then results in the downward propagation of a wetting-front, followed by a slow redistribution of water in the soil profile.

Soils are often far from homogeneous in their structure and hydraulic properties but, for convenience rather than realism, the water flow is often divided into two components, matrix flow and pipe or channel flow. Matrix flow, defined as flow through intergranular pores and small voids, may be divided into downslope and vertical components and may occur under both saturated and unsaturated conditions. Having passed through the soil matrix, water with a long residence time in the mineral soil horizons will usually have undergone considerable chemical modification before contributing to the base flow or stream flow.

Pipe or channel flow occurs through the large voids, some centimetres in diameter, which form open passage-ways through the soil and provide a rapid mechanism for the transmission of subsurface flow down a slope. Such water, if restricted to the peaty surface layers of upland podzols, will have little contact with mineral soils and, if acid, little chance of being neutralized by ion-exchange or the products of mineral weathering. Since water can move rapidly through pipes, root channels, etc. only after the soil becomes saturated, pipe flow will make a significant contribution to stream discharge only after prolonged rain has saturated the upper soil horizons.

4.2 Hydrochemical changes in experimental catchments

4.2.1 INTRODUCTION

The chemistry of soil water and stream water is largely determined by the structure, depth, and chemical composition of the various soil layers through which the incoming rain or snow water percolates, and by the actual pathways which it follows through the soil and which, in turn, determine the residence times during which the various chemical and biological processes can act to modify its chemistry. Short-term changes in stream chemistry are largely determined by changes in the route of the percolate through the soil.

In a complex catchment, in which the topography, soil structure and chemistry, and the flow pathways will usually be far from homogeneous, the water reaching the stream will result from the mixing of several components following different hydrological and chemical pathways. Detailed study of these changes on a plot, let alone on a catchment scale, calls for continuous or frequent measurements of water flow and of all the relevant chemical species in three-dimensional networks. This is a difficult, complex, and expensive task that has been undertaken in only a few locations in a small number of countries.

Such comprehensive experiments were a major feature of the Anglo-Scandinavian programme, and were carried out at the Allt a'Mharcaidh and Loch Ard sites in Scotland, the Birkenes catchment in Norway, and the Svartberget catchment in Sweden. The results, from the Allt a'Mharcaidh catchment in particular, will now be described in some detail to illustrate the main characteristics of the water flow and chemistry during periods of low stream flow and also the large, rapid changes that take place during episodic periods of high flow caused by heavy rainstorms or melting snow.

4.2.2 HYDROCHEMICAL CHANGES IN THE ALLT A'MHARCAIDH CATCHMENT

Site description and instrumentation

The catchment, located in the Cairngorm Mountains, is a hanging valley which drains an area of about 12 km^2, dropping from a height of about 1000 m to 300 m at its outlet into the river Feshie, a tributary of the River Spey. The upland plateau lies above 750 m and rises gently to about 1000 m and is largely snow covered in winter. The valley floor, rising from 300 m to 700 m carries the main Allt a'Mharcaidh stream and is covered by extensive peat deposits and moraine.

The Macaulay Institute has carried out a detailed survey of the geology, soil, and vegetation over 15 units, ranging in area from 10 to 250 ha. The underlying rocks are entirely granite, predominantly biotite granite, with rather little variation in structure and composition. The scree deposits are found mainly on the steep slopes, whilst the alluvial deposits, of recent origin, are confined to the bottom land around the streams. Peat, defined as organic deposits more than 50 cm thick and strongly acid, is confined mainly to the valley floor. The lower slopes of the valley have large stands of native Scots pine on a range of soils, notably iron podzols, peaty podzols, and peat. Ground cover is mainly heather, moss, grasses, and some bracken. On the valley walls, which are generally quite steep and smooth, the dominant vegetation is that of a heather moor, the faces of many slopes being covered in mountain blanket bog. The exposed rocky ridges and the summit have colonies of alpine lichens, fringe-moss, and mountain azalea.

One of the aims of the survey was to locate areas of uniform soils that could be used as experimental plots. These had to be near a stream and representative of the dominant topography and soil in that part of the catchment. The chosen experimental plots are marked P_1, P_2, and P_3 in Fig. 4.1.

Chemical inputs to the catchment were determined from measurements of precipitation both above and below the tree or vegetation canopy, of the stem flow, and by interception of the cloud and fog water. Hourly rainfall totals were measured at six sites, and comprehensive hourly meteorological data provided at two sites. Automatic snow-melt samplers were installed at three sites, and snow-depth surveys were made weekly and more frequently during snow-melt episodes. Chemical analyses of bulk precipitation, snow, and melt-water for pH, alkalinity, total organic carbon (TOC), and all the main ionic species were carried out on a weekly basis, but much more frequently during storms and snow-melt episodes.

The chemical outputs of the catchment were determined from the flow measured in the three streams, together with pH, conductivity, and temperature at 20 minute intervals, and the full stream water chemistry was determined weekly, or every 1–3 hours during episodes.

A series of borehole piezometers, installed along three profiles perpendicular to the main stream and extending some 200 m upslope, were monitored to determine the height of the water table during storm events, and hence the contribution of the shallow ground water to the stream.

The structure and chemistry of the soils and the chemistry and flow of the percolating soil water were determined at the three experimental sites

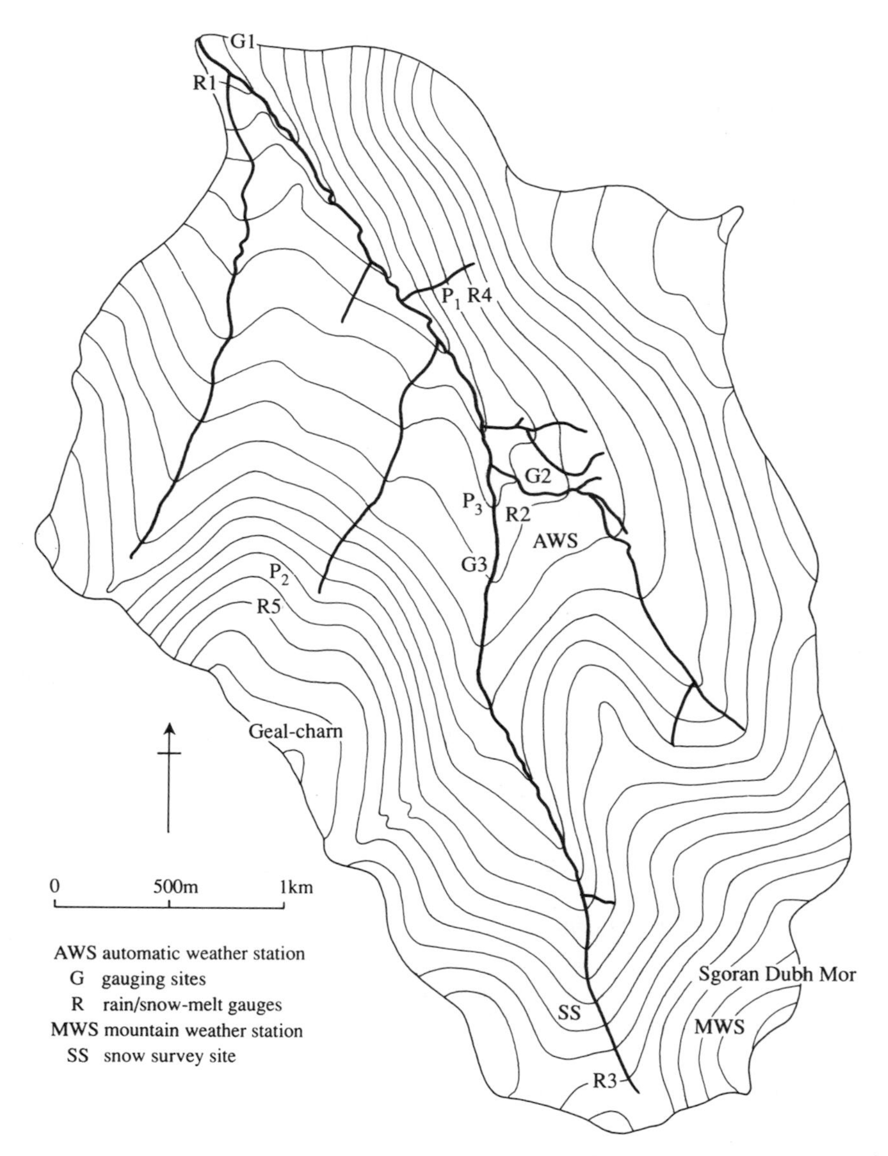

Fig. 4.1 The Allt a'Mharcaidh catchment showing the locations of the experimental plots and the monitoring stations.

on peaty podzol, alpine podzol, and peat, all close to a stream and under upland heath vegetation.

The soils were characterized by particle size and shape, packing, porosity, and crystalline structure. Soils from different horizons were analysed for their major inorganic and organic constituents, for potentially toxic heavy metals, humic substances, organic acids, carbon to nitrogen ratios, etc. Soil moisture was measured at, and soil water extracted from, an array of points in each of three pits dug in each of the three sites. Soil water potential (pressure) was measured by an array of some 20 porous cup tensiometers which provided a three-dimensional representation of soil water pressure from which flow directions were deduced. They also helped to identify areas of saturated, unsaturated, channel, and macropore flow. The soil water was also extracted by simple gutter lysimeters at three different horizon depths in each pit. Water flow from the lysimeters was monitored by tilting bucket collectors, and samples were collected for chemical analysis of all the major ionic species, pH, conductivity, alkalinity, and TOC, every 2 weeks. Concentrations of water-soluble aluminium species were determined by the Driscoll method. Hourly sampling during episodes was facilitated by an automated system installed in one pit on each site. Probes to measure the concentration of CO_2 in the soils were installed on the two podzol sites.

Data from networks of tensiometers and piezometers were used to study the water-flow paths on the plot and hill-slope scales, and to incorporate this understanding into both plot- and catchment-scale models.

Experimental results

During periods of low flow, when the soil is dry, the stream chemistry is dominated by the alkaline base flow fed by the ground water. While the flow remains low, the onset of rain rapidly dilutes the alkaline base flow (the calcium concentration falls), the upper soil horizons having little effect. But, as the rain continues and the flow increases, a more complex mixing process occurs whereby the relative contributions from the different soil horizons, with their differing chemistry, gradually change as the soil profile wets up.

Slow drainage through the soil horizons, leading to lateral flow down the hillslope, is mainly responsible for the ‘humped’ or ‘whaleback’ response in the stream hydrograph shown Fig. 4.2. Lateral flow in the alpine podzols on the upper slopes of the catchment occurs partly in the O/H surface horizons but the bulk of the water penetrates into the highly permeable mineral B/C horizons (see Section 5.1 for definitions and descriptions of the various soil horizons). In the peaty podzols that dominate the lower slopes, flow occurs mainly in the B/C horizons as a result of flow at depth upslope, and by infiltration from the surface. Flow

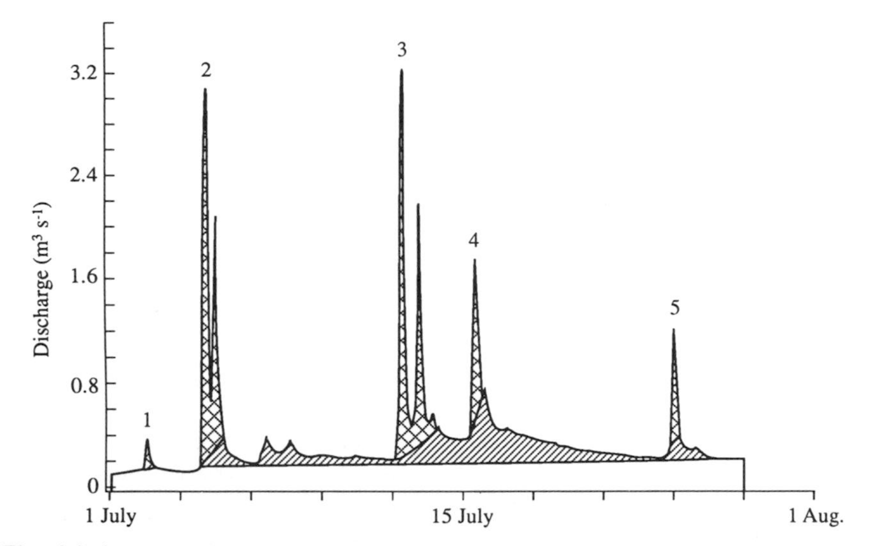

Fig. 4.2 An example of the three modes of flow response at the gauging site G1 (see Fig. 4.1) during July 1988. Cross hatching delineates the spike, hatching represents the 'whaleback', and the base flow is unshaded. (From Jenkins *et al.* 1990, by permission of the Cambridge University Press.)

in the organic surface layers is greater than in the upper parts of the B horizon where it is restricted by impermeable iron pans.

In the areas covered by peat at the base of the slopes, flow occurs mainly below the peat at the interface with the mineral E horizon and is fed from both the upslope peaty podzols and from below by the underlying ground water. However, the peat is characterized by channels and pipes that penetrate to the E horizon so a major contribution to the stream flow comes from the deeper mineral horizons where the acidity is partly neutralized by exchange with Ca^{2+} ions. Aluminium in the organic surface layers exists largely in organic complexes, while inorganic aluminium appears in highest concentrations in the B horizon.

As the rain continues and the water table rises, the flow is confined increasingly to the surface horizons accompanied then by an increase in H^+ concentration and decreases in Ca^{2+} and inorganic aluminium.

In prolonged heavy rain, when the soil profile becomes fully wetted and the surface layers saturated, the pressure exerted by the water upslope tends to force out water stored in the surface layers nearer the valley bottom. The chemistry of this pre-event water, reflecting the base status of the soil in the surface horizons and its ability to adsorb sulphur, is usually characterized by high concentrations of TOC with much of the aluminium being organically complexed.

The sulphate content of the run-off remains rather steady, indicating

Table 4.1 Characterization of Scandinavian SWAP field sites

	Birkenes I	Atna (Storbekken)	Høylandet (Ingabekken)	Svartberget
Area of catchment studied (ha)	0.41	4.35	0.21	0.50
Altitude (m)	205–300	720–1535	280–370	235–310
Bedrock	Granite	Quartzite	Granite	Gneiss, schist
Precipitation (mm)	1500	550	1300	710
Excess sulphate in precipitation ($\mu eq\ l^{-1}$)	55	29	9	40
Approximate pH range of stream	4.2–5.3	4.7–5.5	4.8–6.6	4.2–6.8
Characterization	Acidified	Transitional	Pristine	Transitional
Sulphate deposition (wet and dry) ($gS\ m^{-2}\ yr^{-1}$)	1.7	0.25	0.35 (1/2 sea-spray)	0.6

considerable storage in the soil. In these conditions, rapid flow of the storm water often takes place across the surface, and through the pipes and channels in the peaty podzols on the mid-slopes, and produces the 'quick-flow spikes' on the stream hydrograph shown in Fig. 4.2. The acid pulses in the stream water tend to lag behind the peak flow because of the delay involved in forcing water from the upper layers of the acid soil into the stream channel.

As the rain eases and the flow recedes, the chemistry of the stream, fed by slow drainage through the mineral soils, gradually returns to pre-storm levels.

4.2.3 THE BIRKENES CATCHMENT

In the Birkenes catchment, the emphasis was on intensive studies of episodic events involving detailed plot-scale investigations on a hillslope. This called for sampling and chemical analysis of the precipitation, snow melt-water, soil, soil water, and stream water, together with measurements of stream flow and ground-water levels. Measurements of soil-water pressure from arrays of tensiometers, and of ^{18}O to ^{16}O ratios, provided information on water-flow paths and residence times.

Although during periods of high flow the upper organic soil horizons provide important pathways, about three-quarters of the storm run-off originated from pre-event water already stored in the soil as was also the case at Allt a'Mharcaidh. Here, too, the stream water at high flow shows elevated concentrations of H^+ ions and reduced Ca^{2+} concentrations, see

Fig. 4.3. The response of inorganic aluminium during heavy rains depends on the antecedent conditions. While the soil moisture remains low and the rainwater can percolate to the mineral B horizons containing most of the inorganic aluminium, concentrations of the latter increase in the stream water but, as the soil profile becomes saturated and the water table rises to confine the flow mainly to the surface layers, the concentrations of both inorganic aluminium, Al_i, and Ca^{2+} fall. In fact, the observed concentrations of Ca^{2+} in the soil waters of the O/H and B horizons are too low to account for those in the stream, which suggests that the stream bed may be a source of calcium. It may also be a source of inorganic aluminium mobilized by the acidic flow, notably from aluminium hydroxides precipitated and adsorbed on dense growths of mosses and liverworts on the stream bed.

4.2.4 THE SVARTBERGET CATCHMENT

A field study to determine the pathways and chemistry of the percolate on its way to the stream during heavy rain events was conducted on a hillslope adjacent to an episodically acid stream at Svartberget in northern Sweden. The 50 ha research site slopes at an angle of 5–10 degrees toward the stream and is underlain by 30 m of till derived from gneissic bedrock. Soils on the till are well developed iron podzols blending into humus podzols near the stream. The hillslope is afforested with Norway spruce.

Transects parallel to the downslope were instrumented with piezometers and ground-water tubes to provide cross-sections of hydraulic potential which, together with estimates of hydraulic conductivity, made it possible to deduce flow pathways and flow rates of water through the hillslope. The volume and chemical composition of the run-off were calculated from the measured differences in discharge and chemistry at two weirs in the stream situated one above, and the other below, the instrumented hill slope.

Each of three storms producing more than 20 mm of rain resulted in episodes of increased acidity with no accompanying increase in SO_4^{2-} ions. However, the total organic carbon more than doubled during each episode. The increased acidity was due mainly to organic acids originating from the forest litter. Most of the aluminium was held in organic complexes, only 20 per cent being in inorganic form. During the episodes, the shallow ground water was very alkaline. In the soil water, concentrations of base cations and the alkalinity increased with increasing depth, whereas total aluminium decreased.

Isotopic measurements on the stream flow indicated that 60–80 per cent of the run-off from the storms was pre-event water. Most of this passed between the upper 15–45 cm of the soil during the few days when

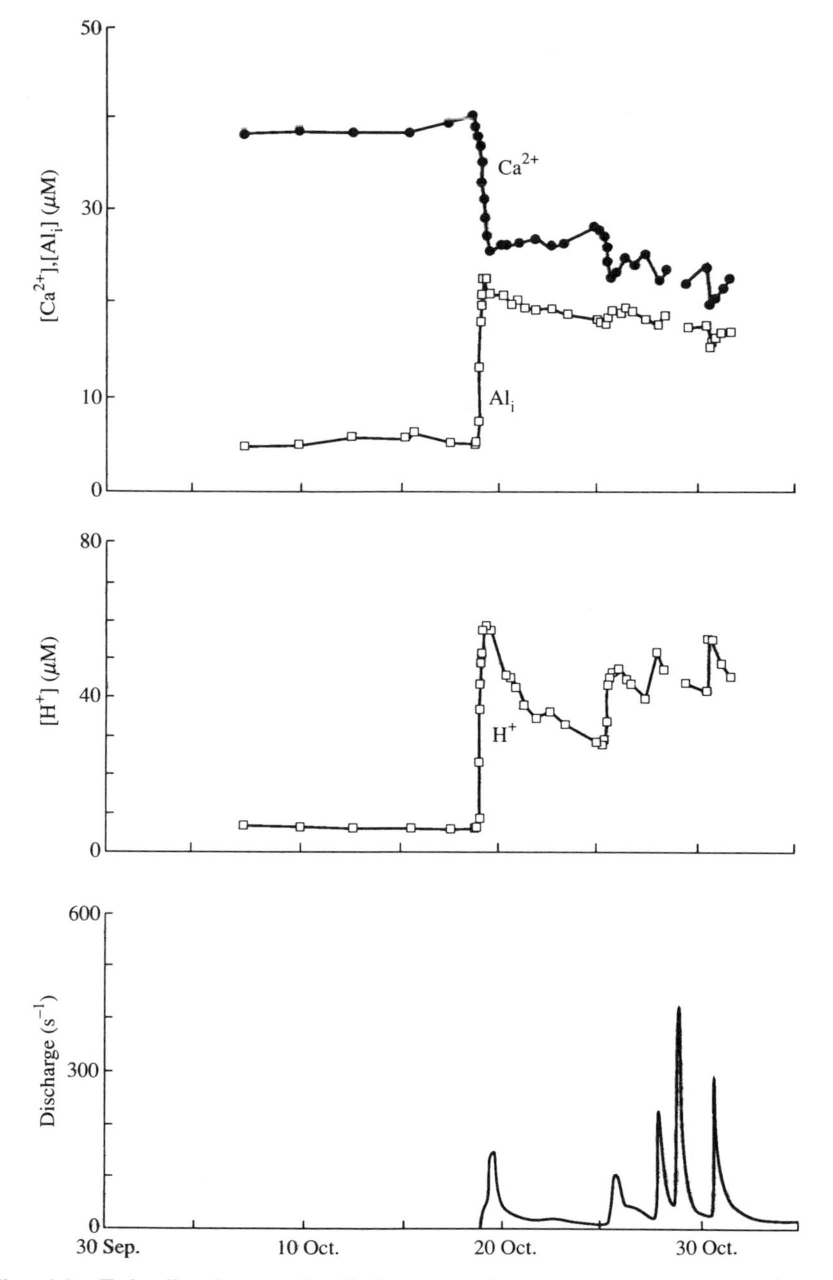

Fig. 4.3 Episodic changes in discharge, and stream water concentrations of Ca^{2+}, Al_i, amd H^+ at the Birkenes catchment during October 1986. (From Christophersen *et al.* 1990, p. 101, by permission of the Cambridge University Press.)

the ground-water levels were elevated by as much as 80 cm by the rain storms. In effect, the rising ground water confined the flow to these upper soil layers, which contained a large reservoir of organically acidified pre-event water. When the rainfall ceased, these surface flow pathways were the first to be drained and thereafter contributed a much smaller proportion of the total flow.

4.2.5 LOCH ARD CATCHMENTS

The afforested Chon and Kelty catchments in the Loch Ard area of west-central Scotland showed some interesting additional features. The flow of soil water is normally greatest at depths below 50 cm, but during heavy rain and snow-melt episodes, the wetting front moves up the soil profile. The leachate from the O horizon is consistently more acidic than that from the B/C horizons and richer in sulphate produced by the mineralization of organic sulphur. Although both catchments receive much the same acidic deposition, the stream water at Chon has a much higher concentration of Ca^{2+} because of a local source, a deposit of dolerite in the upper part of the catchment.

4.2.6 CONCLUSIONS

Although the short-term changes in soil- and stream-water chemistry recorded in the three catchments show some differences, they have many common features. In all cases the changes are strongly dependent on the intensity and duration of the rainfall, which determine the progress of the wetting fronts through the soil profile and hence the pathways and residence times of the percolate in the soil on its way to the stream.

On the evidence from these catchments, the contemporaneous infiltrating rainwater does not play the direct and dominating role in the stream chemistry assumed in most models (see Chapter 9). If the picture of water flow deduced for the SWAP catchments holds generally for catchments of similar topography and soil structure, it will be necessary for models attempting to simulate and predict short-term hydrochemical changes to represent the important effects of preferred (quick flow) pathways, and also the displacement of pre-event water, which damps the response of the stream to fluctuations in the rainfall and its chemistry.

4.3 References

Christophersen, N., *et al.* (1990). In *The surface waters acidification programme*, (ed. B.J. Mason). pp. 97–106. Cambridge University Press.

Jenkins, A., *et al.* (1990). In *The surface waters acidification programme*, (ed. B.J. Mason). pp. 47–55. Cambridge University Press.

FURTHER READING

Bishop, K.H., *et al.* (1990). In *The surface waters acidification programme*, (ed. B.J. Mason). pp. 107–19. Cambridge University Press.

Ferrier, R.C. and Harriman, R. (1990). In *The surface waters acidification programme*, (ed. B.J. Mason). pp. 9–18. Cambridge University Press.

Grip, H. and Bishop, K.H. (1990). In *The surface waters acidification programme*, (ed. B.J. Mason). pp. 75–84. Cambridge University Press.

Harriman, R., *et al.* (1990). In *The surface waters acidification programme*, (ed. B.J. Mason). pp. 31–45. Cambridge University Press.

Seip, H.-M., *et al.* (1990). In *The surface waters acidification programme*, (ed. B.J. Mason). pp. 19–29. Cambridge University Press.

Wheater, H.S., *et al.* (1990). In *The surface waters acidification programme*, (ed. B.J. Mason). pp. 121–35. Cambridge University Press.

5
The role of soils in the acidification of surface waters

5.1 Some notes on soil structure and chemistry

Soil consists of a sequence of chemically and biologically differentiated layers (horizons) formed by the action of chemical, biological, and physical processes on the residues of rocks and minerals, which themselves are the result of the physical and chemical weathering of the original rocks exposed at the surface. In the process of weathering, fracture occurs along planes of weakness in the rocks, and their mineral components (the primary minerals) react chemically with water, oxygen, and carbonic, sulphuric, and nitric acids derived largely from the atmosphere (acidic deposition), with organic acids, and by the action of living organisms in the soil. The continuous operation of these reactions transforms the rocks into soil, the primary minerals being converted into secondary minerals, which are then further degraded by continued weathering (see below).

The combined action of the weathering processes in the soil results in selective vertical transfer and deposition of substances by the percolating water and causes the soil to take on a layered structure, usually in four main horizons designated A, B, and C from the surface downwards in the soil profile, together with the underlying consolidated rock R. These horizons are distinguished by differences in structure, composition, texture, particle size, and colour brought about by the accumulation or decomposition of various materials *in situ* or the release of compounds that may be dispersed, dissolved, or transported from one horizon to another in the percolating soil water. The resulting changes in the physical and chemical environment are often accompanied by changes in the populations of living organisms which, in turn, may influence the further evolution and the properties of the soil.

In the upper, A, horizon, organic matter tends to accumulate, weathering is very active, but the resulting products are partially removed to lower layers. In a rich soil, the A horizon may contain humus to a depth of 15 cm or more. The B horizon, which may be up to 1 metre deep, is a zone of more moderate weathering and tends to accumulate products removed from the A horizon and so is termed the illuvial layer. It may

Table 5.1 Soil horizons in podzols, formed under coniferous forest and heath vegetation in cool, humid climates

A, O, or H	Black pH < 4	Organic or humus layer; peaty material, litter, and incompletely decomposed organic matter with fibrous structure
A1	Dark	Mineral layer near the surface with a good deal of organic matter; clay mineral decomposition products, e.g. chlorite, vermiculite
A2 or E	Ashy grey leached pH ∼ 4	Eluvial horizon underlying O, H, A1; composed mainly of sand and silts; clay minerals, Al and Fe largely removed; low organic content; loss of Fe gives light colour
B	Orange–brown pH 4–5	Illuviated mineral soil with no rock structure; weathering and dissolution of minerals has taken place; Al and Fe translocated from higher levels to give orange–brown colour; may permit considerable sulphate adsorption.
C	pH 5–6	Mineral layer with unconsolidated material; clay minerals (kaolinite, micas, chlorite, some gibbsite) as result of weathering of primary minerals; also quartz, quartz–mica schists, and glacial till.
R		Bedrock — granites, gneisses, schists containing quartz and feldspars (primary minerals).

contain a high proportion of clay, become largely impermeable, and hold 20 cm or so of available water. The C horizon contains the parent materials from which the A and B horizons are formed. A particular soil may not exhibit all four horizons for a number of reasons, for example erosion of the A layer, insufficient time to form a marked B layer, or rapid weathering leaving no parent material to form a C layer.

The main soil horizons just described, and listed in Table 5.1, may be sub-divided and classified as the result of modifications in structure, texture, and colour brought about by chemical reactions and translocation of materials. Thus the topmost surface layer may be composed almost entirely of humus or peaty material and designated an H or O (organic) layer. The A layer, a mineral layer near the surface containing a good deal of organic matter, is sometimes underlain or replaced by an eluvial (E) layer, predominantly of sand or silt, low in organic matter, light in colour, most of its clay, aluminium, and iron having been leached out. In some classifications, hybrid terms such as Ah and Ae are used.

Soils formed under coniferous or heath vegetation in cool, humid climates often have an ashy surface layer resting on a darker, rich stratum. These are termed *podzols* (ashy soils). The ashy, sandy layer is the result of eluviation by the acidified percolate of the sesquioxides Fe_2O_3 and Al_2O_3 and their transport to a lower illuvial layer. Acidification of the percolate is often enhanced by the acid raw humus produced by coniferous tree litter and, to a lesser extent, by the more weakly acid humus produced by deciduous trees and heath-type vegetation.

Because weathering by the acidic percolate is very rapid, amorphous clays of high cation exchange capacity (see Section 5.4.2) are formed in the illuvial B layer and retard acidification of the soil, the pH of which increases with increasing depth. In the ashy surface layer, which consists largely of silicates very resistant to weathering, the pH generally decreases with increasing depth.

At a later stage of development, iron may be deposited in a narrow impermeable band a few millimetres thick at the base of the B horizon. The formation of such an iron pan (placon), a common feature of upland soils receiving high rainfall, may block the normal flow pathways and force the percolate to take a different route. If it confines the lateral flow of water to the surface acid organic (A,O,H) and leached mineral (E) horizons, it may enhance surface water acidification. If the climate and topography are favourable, highly acidic hill peat may form above the iron pan, a final stage in the acidification sequence. The classification, structure and mineral composition of the soil horizons in podzols are summarized in Table 5.1.

In general, the rain falling on such soils will be less acid than the surface layers so that ion exchange will remove H^+ ions from the soil and deposit cations, mostly Ca^{2+}, from the rainwater. However, as the water penetrates into the less acid soil horizons, where the dominant exchangeable cation is often Al^{3+}, H^+ ions are absorbed on the soil particles and Al^{3+} ions are released into the soil solution. All mobile cations must be accompanied by mobile anions to preserve charge balance. The accompanying anion for Al^{3+} is often SO_4^{2-} and the two may be precipitated in the lower soil layers as insoluble, basic aluminium sulphate.

Although the soils of acidified catchments tend to adsorb and retain some of the incoming sulphate, the processes that determine the sulphate balance are complex and not well understood. Whilst adsorption of sulphate ions and their removal from the percolate tends to reduce the rate of leaching of cations, observed rapid increases in stream sulphate during periods of high flow suggest that some of the sulphate stored in the soil is in highly reactive, soluble form. On the other hand, long-term data from the SWAP sites indicate little delay in stream sulphate response to the reduction in sulphate inputs over the past decade — see Fig. 6.1.

5.2 Soil formation by weathering of rocks and minerals

The primary minerals are the constituents of igneous and metamorphic rocks. The former result from the crystallization of silicate melts (magma) and include granite, composed of the minerals quartz (SiO_2), orthoclase feldspar ($KAlSi_3O_8$), and mica or hornblende (both complex silicates), and rocks such as dunite composed almost entirely of olivines (e.g. Mg_2SiO_4 and Fe_2SiO_4).

Quartz constitutes about 25 per cent of igneous rocks and weathers only very slowly.

Feldspars, a family of aluminosilicates, the parents of clay minerals and micas, constitute nearly two thirds of igneous rocks and nearly 50 per cent of rocks exposed on the Earth's surface. They are of two main classes: *K feldspars*, e.g. orthoclase which weathers only slowly by dissolution and hydrolysis, i.e.

$$\begin{aligned} 4KAlSi_3O_8 + 4H_2CO_3 + 18H_2O \rightarrow\ & 4K^+ \text{ available for cation exchange} \\ & + 4HCO_3^- \text{ anions} \\ & + Al_4Si_4O_{10}(OH)_8 \text{ kaolinite} \\ & + 8Si(OH)_4 \text{ silica} \end{aligned}$$

or

$$\begin{aligned} 3KAlSi_3O_8 + 2H^+ + 12H_2O \rightarrow\ & 2K^+ \\ & + KAl_3\,(OH)_2\,Si_3O_{10} \text{ muscovite} \\ & + 6Si(OH)_4) \text{ silica;} \end{aligned}$$

and *plagioclase feldspars*, which weather more rapidly, for example

$$\begin{aligned} \underset{\text{(albite)}}{4NaAlSi_3O_8} + 4H_2CO_3 + 18H_2O \rightarrow\ & 4Na^+ \\ & + 4\ HCO_3^- \\ & + Al_4Si_4O_{10}(OH)_8 \text{ kaolinite.} \end{aligned}$$

Olivines are easily weathered by dissolution in carbonic acid formed from CO_2 existing under enhanced pressure in the soil, thus

$$\begin{aligned} (Fe, Mg)_2SiO_4 + 4H_2CO_3 \rightarrow\ & 2(Mg^+, Fe^+) \text{ available for cation exchange} \\ & + 4\ HCO_3^- \\ & + Si(OH)_4 \text{ silica.} \end{aligned}$$

The secondary minerals, produced by the weathering of primary minerals, and which include the clay minerals and micas, are decomposed further in soil solution and make base cations such as Ca^{2+}, Mg^{2+}, Na^+, K^+, and also Al^{3+} available for exchange with H^+ ions.

Some important reactions are given below:

kaolinite

$$Al_4(OH)_8Si_4O_{10} + 10H_2O \rightarrow \underset{\text{(gibbsite)}}{4Al\,(OH)_3} + \underset{\text{(silica)}}{4Si(OH)_4}$$

and

$$Al(OH)_3 + 3H^+ \rightarrow Al^{3+} + 3H_2O,$$

reactions which consume H^+ ions and release Al^{3+} ions in a process known as exchangeable acidity, although the cation exchange capacity of gibbsite (alumina) is quite low;

imogolite

$$Al_2O_3SiO_2 \cdot 2H_2O + 3H_2O \rightarrow 2Al(OH)_3 + Si(OH)_4;$$

muscovite is resistant to weathering but decomposes slowly, i.e.

$$KAl_2(OH)_2\,Si_3AlO_{10} + 10H^+ \rightarrow K^+ + 3Al^{3+} + 3Si(OH)_4,$$

so consuming H^+ ions and releasing K^+ and Al^{3+} but at a rate which is very dependent on pH;

biotite

$K(Mg, Fe)_3AlSi_3O_{10}(OH, F)_2$ decomposes to release K^+, Mg^{2+} or Fe^{2+} and Al^{3+} ions, i.e.

$$KMg_3AlSi_3O_{10}(OH)_2 + 10H^+ \rightarrow K^+ + 3Mg^{2+} + Al^{3+} + 3Si(OH)_4$$

and has a quite high cation exchange capacity; biotite also decomposes to form either

$Mg_3(OH)_2AlSi_3O_{10}$	chlorite,
$Mg_3(OH)_2\,(Si_4\,O_{10}) \cdot xH_2O$	vermiculite, or
$(OH)_4Si_8O_{20} \cdot xH_2O$	smectite,

all of which, in turn, can decompose almost completely, leaving little trace of their decomposition products.

Although the weathering of rocks and the formation of soils occur largely by acid hydroloysis of aluminosilicates, oxidation and direct solution can also be significant, especially the dissolution of limestone by dilute carbonic acid, produced by CO_2 existing under enhanced pressure in the soil, i.e.

$$CO_2 + H_2O \rightarrow H_2CO_3 \xrightarrow{\text{partial}} H^+ + HCO_3^-$$
$$CaCO_3 + H_2CO_3 \rightarrow Ca^{2+} + 2\,HCO_3^-,$$

and also by the simple solution of gypsum

$$CaSO_4 \cdot 2H_2O \rightarrow Ca^{2+} + SO_4^{2-} + 2H_2O.$$

These reactions occur in easily weathered, calcareous soils with pH values often exceeding 7.

5.3 Measurements of weathering rates

Since chemical weathering provides the main long-term sink for H^+ ions, knowledge of weathering rates is essential for long-term predictions of catchment acidification. Several methods have been used to measure to estimate weathering rates within SWAP catchments and with considerable success.

Long-term (historical) rates are determined from the abundance of the various base cations and acid anions from an elemental analysis of the rocks and soils in isotopically dated soil profiles using the accumulation of conserved elements such as zirconium, titanium, and quartz as an internal standard. Current rates are determined from measured inputs and outputs of base cations for the catchment, the former by deposition and the latter by the streams. The calculation assumes a steady state, with negligible depletion of the stores of exchangeable cations in the soil, an assumption that will not always be valid in areas of heavy, increasing acid deposition or of active afforestation.

Weathering rates obtained from chemical analyses of the soil profile often show a considerable spread because they are, in effect, 'point' samples and may not be representative of the catchment as a whole. Mean values are shown in Table 5.2. The budget method gives current values averaged for the whole catchment.

In general, there is fair agreement between the different measurements (major anomalies can usually be explained in terms of special features of the mineralogy or water flow), so it is now possible to make useful estimates of weathering rates for several different types of catchment based on a few key parameters such as mineralogy, soil texture, and soil depth.

The relative amounts of base cations, Ca^{2+}, Mg^{2+}, K^+, Na^+, lost from a catchment reflect the abundance of these elements in the parent materials of the soils. The average loss rates of base cations from the most acid-sensitive catchments in southern Scandinavia are 5–20 meq m^{-2} yr^{-1}, depending on the mineralogy. Areas of glacial till derived from acid granites and gneisses, with a soil depth of 1–5 m, have rates of 20–30 meq m^{-2} yr^{-1}. Rates of 25–75 are found in Scottish catchments and as high as 120 meq m^{-2} yr^{-1} in the pristine Høylandet catchment, where the release of Ca^{2+} and Mg^{2+} is particularly high because of the basic nature of the bedrock. In the forest soil at Svartberget, the rate was 46 meq m^{-2} yr^{-1} in the top 20 cm of the mineral horizons — about

Table 5.2 Sulphate deposition and soil weathering rates for catchments

Catchment	SO_4 deposition	Rate (meq m^{-2} yr^{-1})	Mean weathering rate (meq m^{-2} yr^{-1})	Method	Dominant soil type	Average pH of stream
Høylandet	L	21	124	LT	Podzols, peat	5.1
Sogndal	L	15	56 16 31	LT C LT	Thin soil on gneiss	—
Svartberget	L, M	36	85 150	C Sr	Glacial till	4.7
Mharcaidh	M	40	45 49	LT C	Podzols, peat	6.5
L. Chon	H	135	72 44 34	LT C Sr	Podzols, gleys	5.1
L. Kelty	H	230	26 19	LT C	Gleys, peat	4.5
Risdalsheia	H	100	4	LT	thin organic soil on granite	4.0

L low, M medium, H high. LT long-term historical estimates, C current rates from budget studies, Sr current rates from strontium isotope measurements.

three times higher than in the next 60 cm, probably owing to the lower pH caused by organic acids and higher water flow.

In a chronosequence of seven profiles aged from 80–13 000 years BP (before present), from Glen Feshie, Scotland, the calculated weathering rate decreases exponentially with time from 70 to less than 5 meq m^{-2} yr^{-1} after 2000 years. The initially rapid rates are probably due to the exposure of fresh mineral surfaces and the presence of fine particles in the recently laid down sediments. After the highly reactive particles have been depleted, the weathering approaches a steady state.

Reference to Table 5.2 indicates that the low acidification of surface waters at Høylandet and Sogndal, high acidification at Kelty and Gårdsjøn, and intermediate values at Allt a'Mharcaidh, can be accounted for by the ratio of the rates of acid deposition and weathering, the anomalies at Svartberget and Chon being readily explained by the dominant influence of organic acids at Svartberget and the presence of a local source of Ca^{2+} and Mg^{2+} at Chon.

At Kelty Water, where the soil is a peaty gley having a base cation exchange capacity (CEC) of typically 4 eq kg^{-1}, mostly confined to the top metre, a weathering rate of 20 meq m^{-2} yr^{-1} indicates a time constant for exponential depletion of basic cations of about 200 years. For the peaty podzols of Allt a'Mharcaidh, where the CEC is only 1.5 eq kg^{-1} and the weathering rate rather higher at 50 meq m^{-2} yr^{-1}, the depletion time constant is only about 30 years. For the L. Chon catchment, with its local doleritic source of base cations, the corresponding time is about 100 years.

5.4 Soil–water interactions

5.4.1 INTRODUCTION — SOME BASIC CONCEPTS

The degree of acidification of soil and soil water is determined not only by the acidic inputs from deposition and the chemical properties of the soil, but also by the water-flow pathways through the soil, which determine the residence time and hence the time available for the chemical and biological processes to operate. Several such processes modify the chemistry of the infiltrating water as it percolates through the various soil horizons, some tending to acidify it, whilst others have a neutralizing effect.

The acidity, alkalinity, and buffering of soil solutions

An *acid* is, by definition, an aqueous solution with an excess of protons. Accordingly, it acts as a donor of protons (or an acceptor of OH^- ions) in a chemical reaction that tends to neutralize the solution. The standard measure of the acidity of an aqueous solution is its pH, equal to $-\log[H^+]$, where $[H^+]$ is the molar concentration of H^+ ions (protons). Pure neutral water is weakly dissociated with $[H^+] = 10^{-7}$ M and pH 7. Strong acids such as H_2SO_4, HNO_3, and HCl are completely ionized in solution. Weak acids such as carbonic acid (H_2CO_3) and acetic acid (CH_3COOH) are only partially ionized. Water in equilibrium with atmospheric carbon dioxide is, in effect, very dilute carbonic acid with pH 5.6.

A *base* is a proton acceptor (or OH^- donor). Strong bases are all ionic metallic hydroxides, e.g. NaOH, KOH, LiOH, $Ca(OH)_2$, and $Mg(OH)_2$, and become completely ionized in solution. Weak bases include hydroxides of less reactive metals, e.g. $Al(OH)_3$ and $Fe(OH)_3$, and are only partially ionized in solution. A base can *neutralize* an acid to produce a salt plus water, for example

$$Al(OH)_3 + H_2SO_4 \rightarrow AlOHSO_4 + 2H_2O.$$

The *base neutralizing capacity* (BNC) of a soil solution is defined as

$$\text{BNC} = [H^+] + [H^+]_{org} + [H_2CO_3] + 3[Al^{3+}] + 2[AlOH^{2+}] + [Al(OH)_2{}^+] - [OH^-],$$

the first three terms representing the concentrations of proton donors and the next three terms representing the OH^- acceptors. Conversely, the *acid neutralizing capacity* (ANC) or *alkalinity* of a solution is a measure of its proton deficiency or capacity to accept protons. Thus

$$\text{ANC} = [HCO_3{}^-] + 2[CO_3{}^{2-}] + [OH^-] + [org^-] - 3[Al^{3+}] - 2[AlOH^{2+}] - [Al(OH)_2{}^+] - [H^+]$$

i.e. the stoichiometrically weighted sum of the concentrations of all the proton acceptors minus the concentrations of all the OH^- acceptors and proton donors.

Buffering and neutralization

A buffered solution is one that can accept additional H^+ or OH^- ions without changing its pH. Buffering against the addition of H^+ ions requires the presence of a strong base and a weak acid and the salt of a weak acid, e.g: $Ca(OH)_2 + H_2CO_3 + CaCO_3$, to which added H^+ yields

$$CaCO_3 + 2H^+ \rightarrow Ca^{2+} + H_2CO_3,$$

whereby $CaCO_3$ accepts H^+ ions from solution and releases Ca^{2+} ions with no net increase in H^+. Similarly, a solution buffered against the addition of OH^- ions requires a strong acid and a weak base and a salt of a weak base, e.g. $H_2SO_4 + 2NH_4OH + (NH_4)_2SO_4$, to which added OH^- yields

$$(NH_4)_2SO_4 + 2OH^- \rightarrow 2NH_4OH + SO_4^{2-}$$

with no net increase in OH^-. Aluminium hydroxides such as $Al(OH)_3$ may buffer a soil solution against the addition of H^+ ions, at $pH < 5$, i.e.

$$Al(OH)_3 + 3H^+ \rightarrow Al^{3+} + 3H_2O,$$

releasing Al^{3+} ions to act as a buffer against the addition of H^+.

Buffering of a solution against further acidification or alkalization is often confused with neutralization. It differs in that it involves ion exchange in a reaction rather than direct recombination of H^+- and OH^--bearing ions.

Natural waters will not usually contain sufficient base cations to act as a complete buffer against acid deposition but, even so, the increase in acidity will usually be considerably less than if the same quantity of acid were added to pure water.

5.4.2 THE ROLE OF WEATHERING AND CATION EXCHANGE IN THE ACIDIFICATION OF SOIL AND SOIL WATER

Weathering of mineral rocks and soils, especially of aluminosilicates, releases base cations Mg^{2+}, Ca^{2+}, Na^+, K^+, and also Al^{3+} ions. The surfaces of some mineral soils, notably the clay minerals, expose permanent negative charges and adsorb a compensating layer of base cations, which they may exchange for H^+ ions from the soil solution. Organic matter also adsorbs negative ions produced by the dissociation of weak humic acids and can therefore adsorb H^+ ions from the soil water. The exchange of H^+ ions from the acidic soil solution for base cations from the soil surface tends to neutralize the former, and to acidify the soil unless H^+ ions are consumed in the weathering process (see Section 5.2). Exchange of H^+ ions for Al^{3+}, $AlOH^{2+}$, or $Al(OH)_2^+$ ions does not have a neutralizing effect as the latter are OH^- acceptors, equivalent to proton donors in the solution, so this process is termed 'exchange acidity'.

The capacity of the soil to adsorb base cations and aluminium cations and to exchange these for H^+ ions is called the cation exchange capacity, CEC = $[AlX_3] + [CaX_2] + [NaX] + [KX]$, where [ ] denotes the equivalent concentrations of adsorbed cations (cation-exchange sites), usually expressed in milliequivalents per 100 grams of soil (meq/100 g), for example a CEC of 100 meq Ca/100 g would be equivalent to $0.1 \times 40 = 4$ per cent of calcium by weight. The CEC of clays varies between 5 and 150 meq/100 g, whereas that of organic matter is 150–300 meq/100 g.

The acid neutralizing capacity (ANC) of the soil is measured by the base saturation (BS), the sum of the equivalent fractions of the sites occupied by base cations, the remaining fraction of sites is occupied by Al^{3+} or H^+ ions and is known as the exchangeable acidity. Thus

$$BS = \frac{[CaX_2] + [MgX_2] + [NaX] + [KX]}{CEC} = 1 - \frac{[AlX_3]}{CEC},$$

so that for very high values of CEC, BS tends to unity. Soils of high CEC and BS are able to extract H^+ ions from the soil solution and raise its pH. The acidity of the soil, defined as the decrease in its base saturation, is correspondingly increased.

During the exchange of H^+ ions for base cations on the soil surfaces, the departing base cations must be accompanied by mobile anions, such as SO_4^{2-} ions, in order to maintain the charge balance in the soil water. However, some soils are able to adsorb anions such as SO_4^{2-}, fixed perhaps by Fe^{2+} or Al^{3+} ions to form $AlOHSO_4$ or $FeSO_4$, and so are not available to accompany the base cations into solution. A likely reaction at the soil–water interface is

$$(AlOH)^{2+} + SO_4^{2-} + 2H^+ \rightarrow AlOHSO_4 + 2OH^- + 2H^+,$$

the OH^- and H^+ ions combining at the surface for release into the solution but with no release of base cations. In other words, anion adsorption leads to retention of base cations and retards acidification of the soil, but increases that of the soil water.

In general, neutralization or buffering of the percolate proceeds more rapidly by cation exchange than by mineral weathering, which provides long-term buffering. All the major buffering mechanisms are pH dependent, as indicated below.

Mechanism	Most effective pH range	Fate of H^+ ions
Carbonate dissolution	6.2–8.0	Consumed by dissolution of $CaCO_3$
Aluminosilicate weathering	5.0–6.2	Consumed in weathering process
Cation exchange	4.2–5.0	Exchanged for base cations
Aluminium exchanged acidity	3.0–4.2	Exchanged for Al^{3+} ions from polymeric Al compounds
Iron exchanged acidity	< 3.5	Exchanged for Fe^{2+} ions released by iron oxides

As the dilute acid solution percolates through the soil, chemical weathering proceeds. H^+ ions are consumed, exchanged, or neutralized and so weathering slows down unless there is a continuing supply of H^+ ions originating from, for example, continued acid deposition, organic acids or dissolution of CO_2 produced under enhanced pressure from microbial or root respiration. In the latter case, weathering proceeds until it achieves equilibrium appropriate to the pH of the carbonic acid system.

Waters draining from mineral soils often have pH > 7, not only because of neutralization by the products of weathering, but because of outgassing of dissolved carbon dioxide.

As an example, we may consider the weathering of muscovite in water according to

$$KAl_2(AlSi_3O_{10})(OH)_2 + 10H^+ \rightarrow K^+ + 3Al^{3+} + 3Si(OH)_4$$

for which the equilibrium equation is

$$[K^+][Al^{3+}]^3[Si(OH)_4]^3 = 10^{13.44}[H^+]^{10};$$

now

$$3[K^+] = [Al^{3+}] = [Si(OH)_4],$$

so

$$\log[K^+] = 1.51 - 1.43\ \mathrm{pH},$$

implying that the reaction is strongly pH dependent and that the release of K^+ and Al^{3+} ions by decomposition of muscovite should increase

27-fold for a one unit fall in pH. At pH 7.0, $[Al^{3+}] = 10^{-8}$ M, but the higher concentrations predicted by the above equation at low pH are not likely to be realized because $Al(OH)_3$ is precipitated at pH 4 when $\log[Al^{3+}] = -3.8$. Furthermore, the kinetics of mineral weathering are so slow that thermodynamic equilibrium may not be reached in field conditions when the water chemistry is changing rapidly.

5.4.3 OTHER MECHANISMS INVOLVED IN THE ACIDIFICATION OF SOIL AND SOIL WATER

Whilst weathering of minerals and cation exchange are, in general, of prime importance in modifying the chemistry of the acidified percolate, several other mechanisms, some of a biological nature, may play a role and to differing degrees in different soils. A number of natural processes, and other processes stimulated by cultivation of the land, tend to acidify the soil and would do so in humid climates, even if the precipitation were alkaline.

Respiration is perhaps the most important natural soil acidification process. Soil microbes and plant roots release carbon dioxide which dissolves in water to form carbonic acid and this partially dissociates into H^+ and bicarbonate ions, i.e.

$$CO_2 + H_2O \rightarrow H_2CO_3 \xrightarrow{\text{partial}} H^+ + HCO_3^-.$$

Hydrogen ions are exchanged for base cations and the bicarbonate anions accompany the H^+ ions out of the soil. As the soil pH declines, less carbonic acid dissociates, so at $pH < 4.5$ very little bicarbonate becomes available for leaching and the acidification process becomes self-limiting.

Another important mechanism of soil acidification is the preferential uptake of base cations by growing trees and plants. The roots take up Ca^{2+}, Mg^{2+}, ... from the soil, but release H^+ ions into the soil in order to maintain charge balance. The base cations are partly replenished when they are leached out of the leaves by rain and returned to the soil, and also when the plants decay and die, but if the trees and crops are harvested, the base cations are lost and acidification is permanent.

Ammonium ions deposited from the atmosphere are, to a large extent, exchanged for H^+ ions at the root surfaces of plants, any excess of ammonium being oxidized (nitrified) thus increasing the acidity of the soil:

$$NH_4^+ + 2O_2 \rightarrow NO_3^- + H_2O + 2H^+$$

Organic acids are exuded by soil organisms and are formed during the microbial decomposition of plant litter. These humic or fulvic acids, though weak compared with mineral acids, are stronger than carbonic acid and can therefore acidify the soil to a pH lower than 4.5, and account for the fact that acid soils are very common, especially in forests. Some organic acids are rapidly metabolized within the soil, others form complexes with metallic ions in the upper layers of the soil. This is of particular importance in the case of aluminium, the inorganic species of which are toxic to fish but the organic complexes not.

Other acidifying processes, not involving plants, are the oxidation of sulphur compounds in dry soils, i.e.

$$H_2S + 2O_2 \rightarrow SO_4^{2-} + 2H^+,$$

and the deposition of sea-salts during which Na^+ ions are exchanged for base cations on the soil particles, thereby lowering their cation-exchange capacity.

5.4.4 PROCESSES TENDING TO NEUTRALIZE SOIL ACIDIFICATION

Not all soils are acid because there are a number of acid-consuming processes at work that counteract or at least slow down the acidifying processes. As we have seen, the most important acid-consuming process is chemical weathering of rocks and soils which have a huge capacity for consuming H^+ ions, often equivalent to thousands of years of acid precipitation. However, it is the *rate* of weathering which is the limiting factor, and in acidified catchments this is unable to keep pace with the acid deposition.

Another important acid-consuming process is the adsorption and reduction of sulphate, which is thought to take place largely on aluminium and iron oxides in the lower horizons of some soils. The reaction may be written:

$$RAl_2(OH)_2 + SO_4^{2-} + 2H^+ \rightarrow RAl_2SO_4 + 2H_2O,$$

where $RAl_2(OH)_2$ represents the aluminium or iron oxide. Sulphate adsorption counters soil acidification by removing both the mobile sulphate ion and the hydrogen ion.

The microbial reduction of nitrate (denitrification) and of sulphate,

$$2NO_3^- + 10H^+ \rightarrow N_2 + 4H_2O + 2OH^-$$

and

$$SO_4^{2-} + 4H_2 + H^+ \rightarrow HS^- + 4H_2O,$$

has a similar effect.

Furthermore, plants, in taking up anions such as NO_3^- in nutrients from the soil, release in exchange OH^- and HCO_3^- ions and these tend to neutralize acidity.

The chemical composition of the water entering streams and lakes reflects the net effect of these various acidifying and neutralizing processes, some of which may act synergistically and others in contention during different stages of the journey of the water through the soil. Although it is possible, by comprehensive sampling and analysis, to assess the relative contributions of the various processes in a particular catchment, they will differ in different catchments. This makes it very difficult to devize models of sufficient general validity to predict, in a given case, the hydrochemistry of the emerging water, even if the inputs are adequately determined (see Chapter 9).

5.4.5 THE EFFECTS OF FORESTS, VEGETATION, AND LAND USE

Changes in the chemistry of soils and surface waters may also be induced by changes in land use, afforestation or deforestation, drainage of soils, burning of heather and the surface humus, the use of fertilisers, liming, etc.

As we shall see in Chapter 7, Rosenqvist has consistently argued that the acidification of many catchments in southern Scandinavia was brought about by changes in agricultural practices and by burning of vegetation, rather than by anthropogenic acid deposition. Apart from studies of the interaction between vegetation and acidic deposition, of the differences between the chemistry of throughfall, stemflow, and the incident precipitation described in Chapter 6, little systematic research on the effects of other land-use changes has been carried out. There is, however, some evidence, not always very strong, but suggestive, for the following.

Afforestation, i.e. the planting of coniferous forest or forest regeneration may, over a few decades, produce acidic, organic surface layers due to leaf litter and the enhanced acidity of throughfall and stem flow. In time, the underlying mineral soils may also become acidified, especially if they are low in base cations. On the other hand, trees and growing vegetation take up base cations from the soil through their roots and some of these are leached out of the leaf surfaces by the rain.

Forest clearance may lead to soil erosion, especially on upland slopes, where the rain and snow-melt will tend to run off the surface, making little contact with neutralizing soil. The shallow soil will also acidify and become saturated with sulphate more quickly. On the other hand, there will be reduced trapping of acid aerosols and marine salts and less cation uptake through the roots and this will help to raise the base saturation.

Improved drainage of forests, by the digging of ditches and deep ploughing, may break up iron pans and allow the drainage water to penetrate to the mineral horizons and become partially neutralized. However, oxidation of sulphur and iron in wet soils may cause acidification of the drainage water.

Reduced *animal grazing* leads to the spread of heather on moorland and forest growth on shallow soils, and hence the acidification of surface waters. Conversely, *burning of forest and heather* has the opposite effect and the base cations in the residual ash may have a liming effect on the soil. If the humus layers of the soil are also burnt, the effect will be enhanced. If a crop is not harvested, base cations taken up by growing plants are returned to the soil, in due course, in plant litter. *Crop removal*, however, removes base cations and contributes to soil acidification, unless these are replaced by liming.

The *use of fertilizers* based on ammonia leads to acidification of soils and drainage waters. For example, we may cite oxidation of ammonium sulphate

$$(NH_4)_2SO_4 + 4O_2 \rightarrow 2HNO_3 + H_2SO_4 + 2H_2O,$$

and oxidation of urea and anhydrous ammonia

$$(NH_2)_2CO + 4O_2 \rightarrow 2NO_3^- + 2H^+ + H_2CO_3,$$

and

$$NH_3 + 2O_2 \rightarrow NO_3^- + H^+ + H_2O.$$

Typical inputs of acid from fertiliser ammonia often greatly exceed the atmospheric deposition of nitric and sulphuric acid. In order to maintain the base status of the soil, it would be necessary to apply quantities of *lime* to neutralize the following:

1 kg urea-N requires	3.6 kg $CaCO_3$
1 kg NH_3 requires	3.6 kg $CaCO_3$
1 kg $(NH_4)_2SO_4$ requires	7.2 kg $CaCO_3$

To neutralize 1000 mm of rain of pH 4 would require 50 kg of lime per hectare, but much larger applications would be necessary to achieve the required rate of release of Ca^{2+} ions over a period of several years. The results of some trials on the artificial liming of lakes are briefly described in Chapter 10.

5.4.6 EVIDENCE FOR SOIL ACIDIFICATION

A number of studies have compared present-day measurements of soil acidity with those made many years ago. Some have found little or no change, others have found a significant increase in acidity. One of the most convincing of the latter was carried out in an experimental forest in south-west Sweden. In 1927, O. Tamm measured the pH of 157 soil profiles, carefully noting the position of each one. In 1984, his son, C.O. Tamm, was able to locate these positions within one metre and remeasure the pH

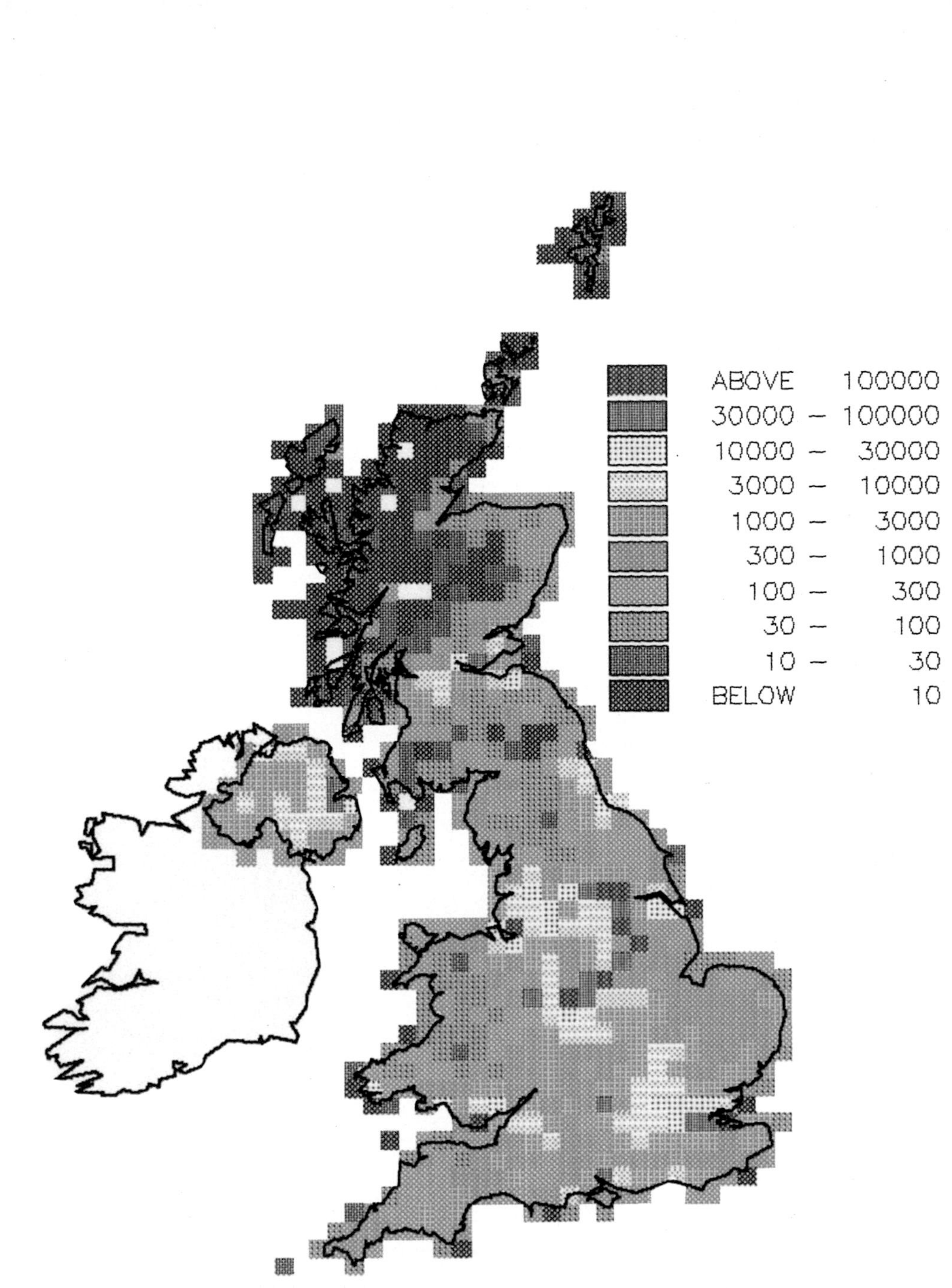

Plate 1 Total sulphur emissions, within the UK for 1987, in metric tonnes. (Figures reproduced courtesy of the Warren Spring Laboratory, Stevenage.)

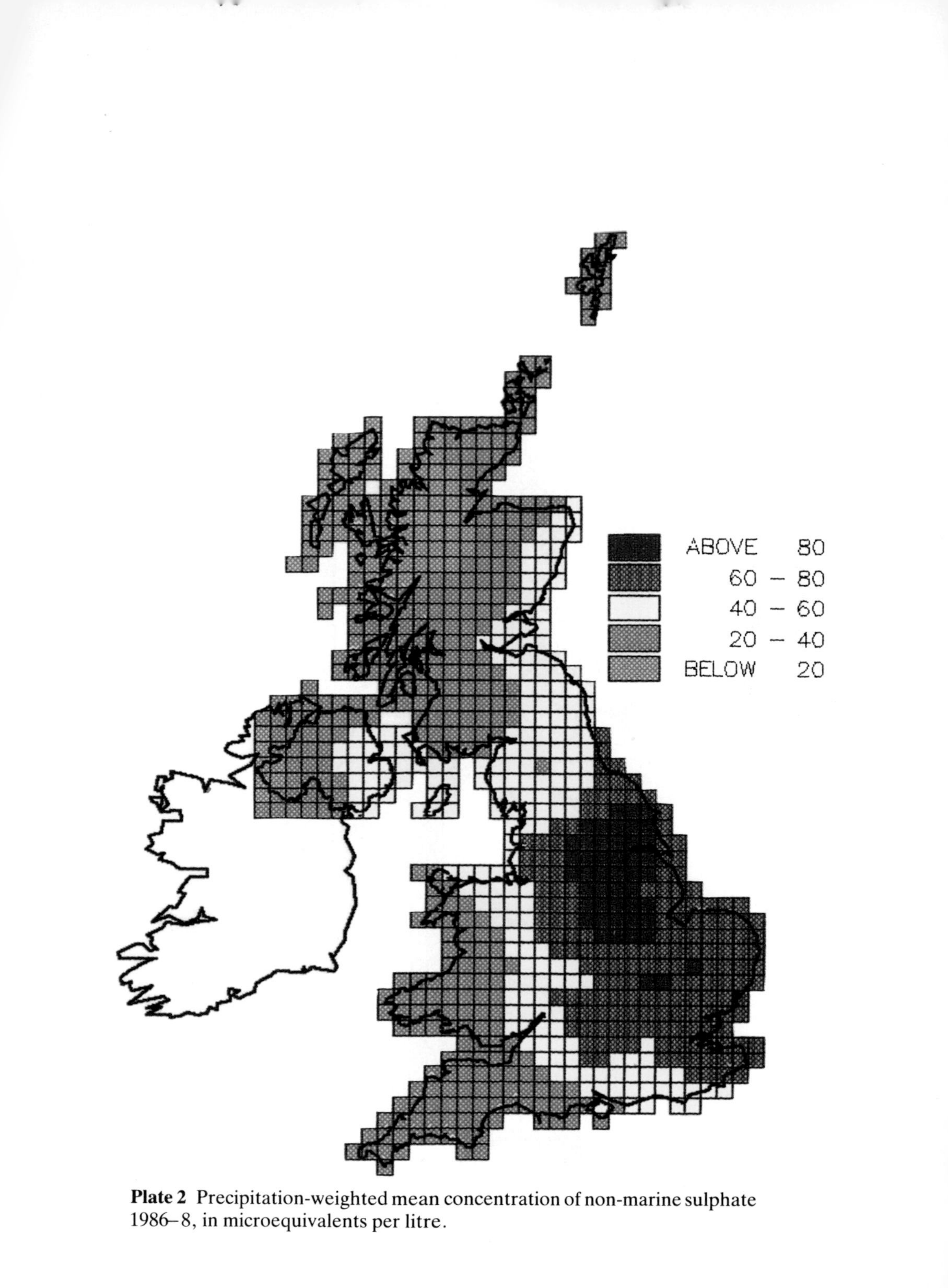

Plate 2 Precipitation-weighted mean concentration of non-marine sulphate 1986–8, in microequivalents per litre.

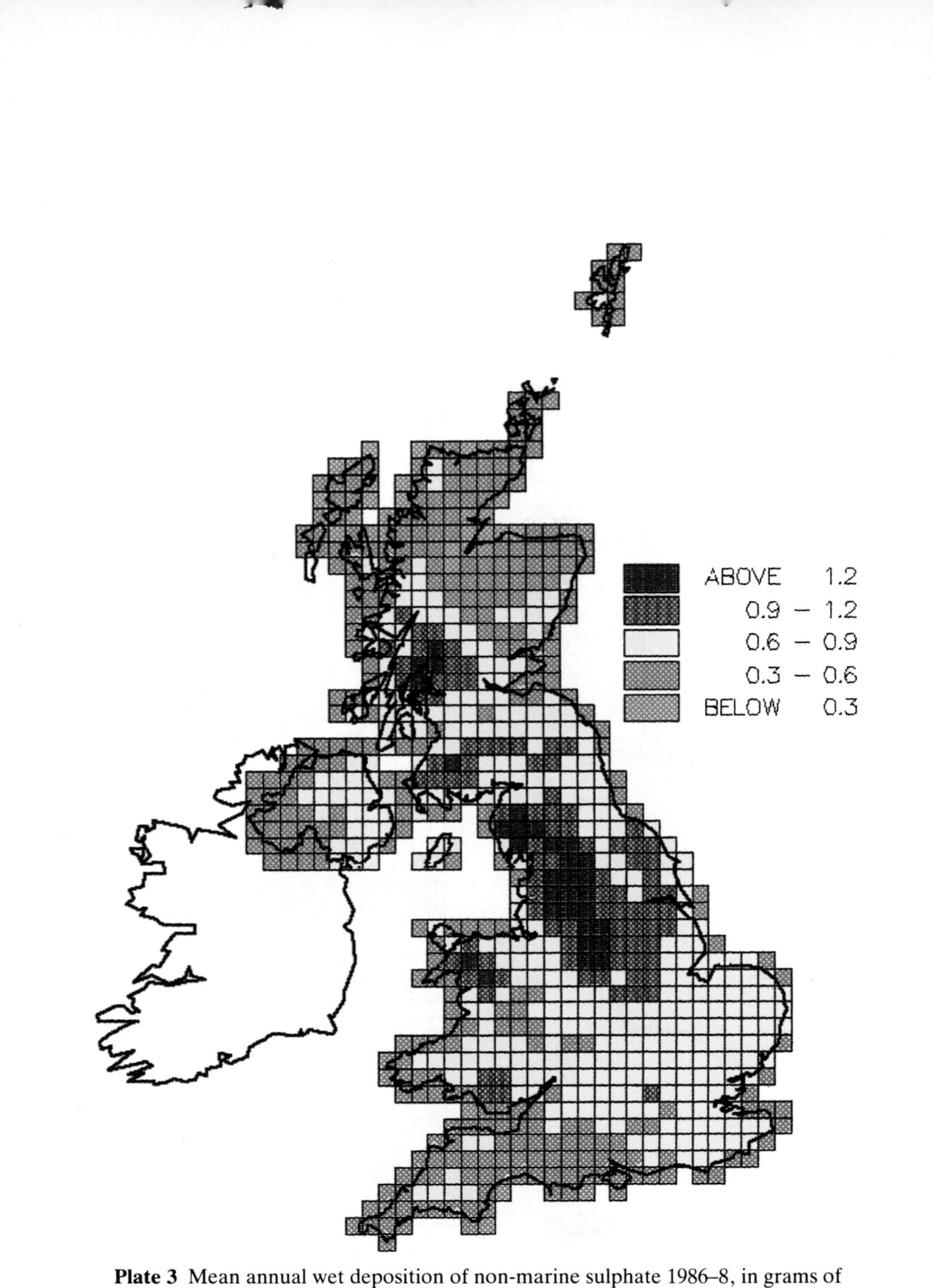

Plate 3 Mean annual wet deposition of non-marine sulphate 1986–8, in grams of sulphur per square metre.

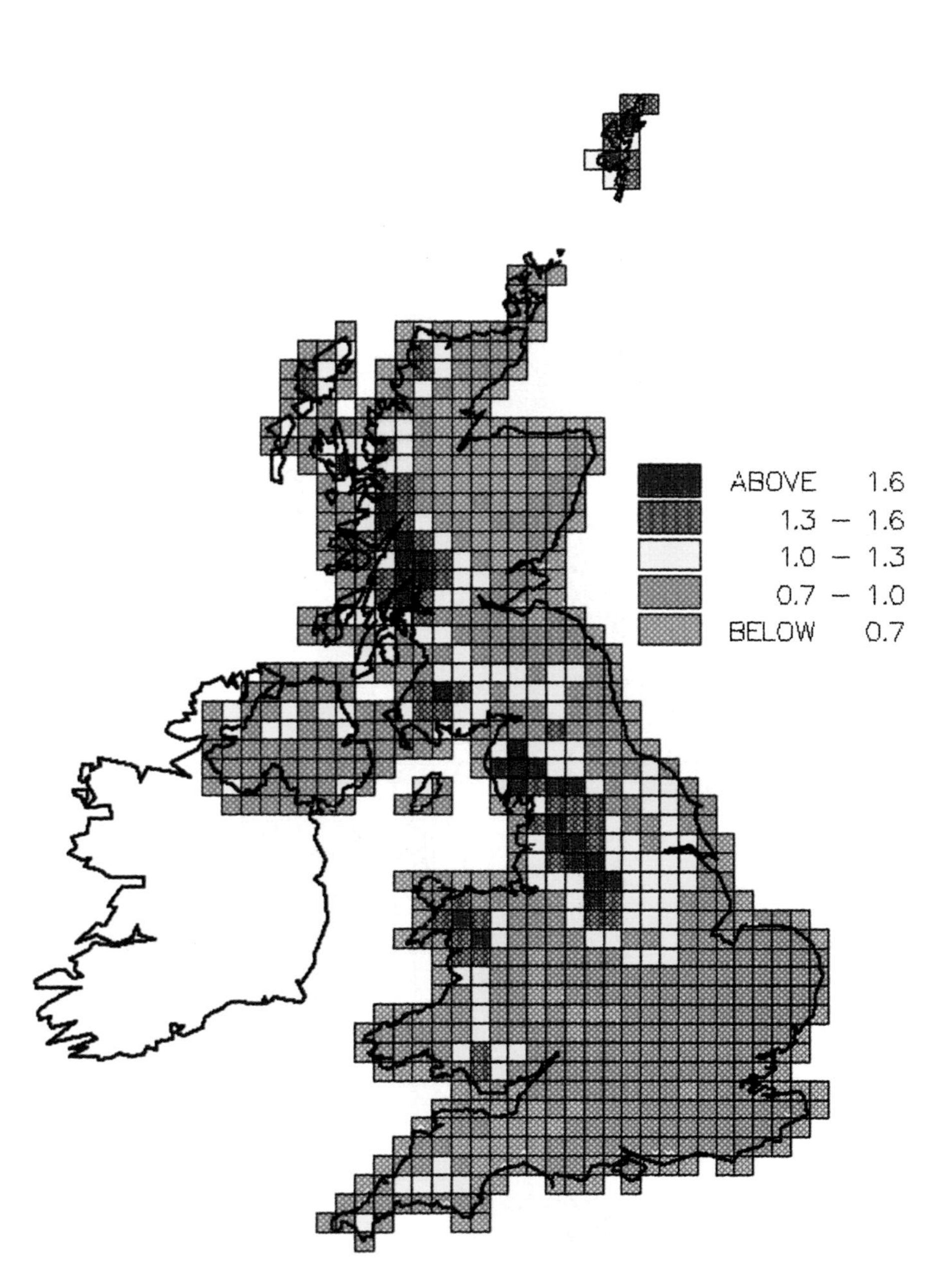

Plate 4 Mean annual wet deposition of total sulphate 1986–8, in grams of sulphur per square metre.

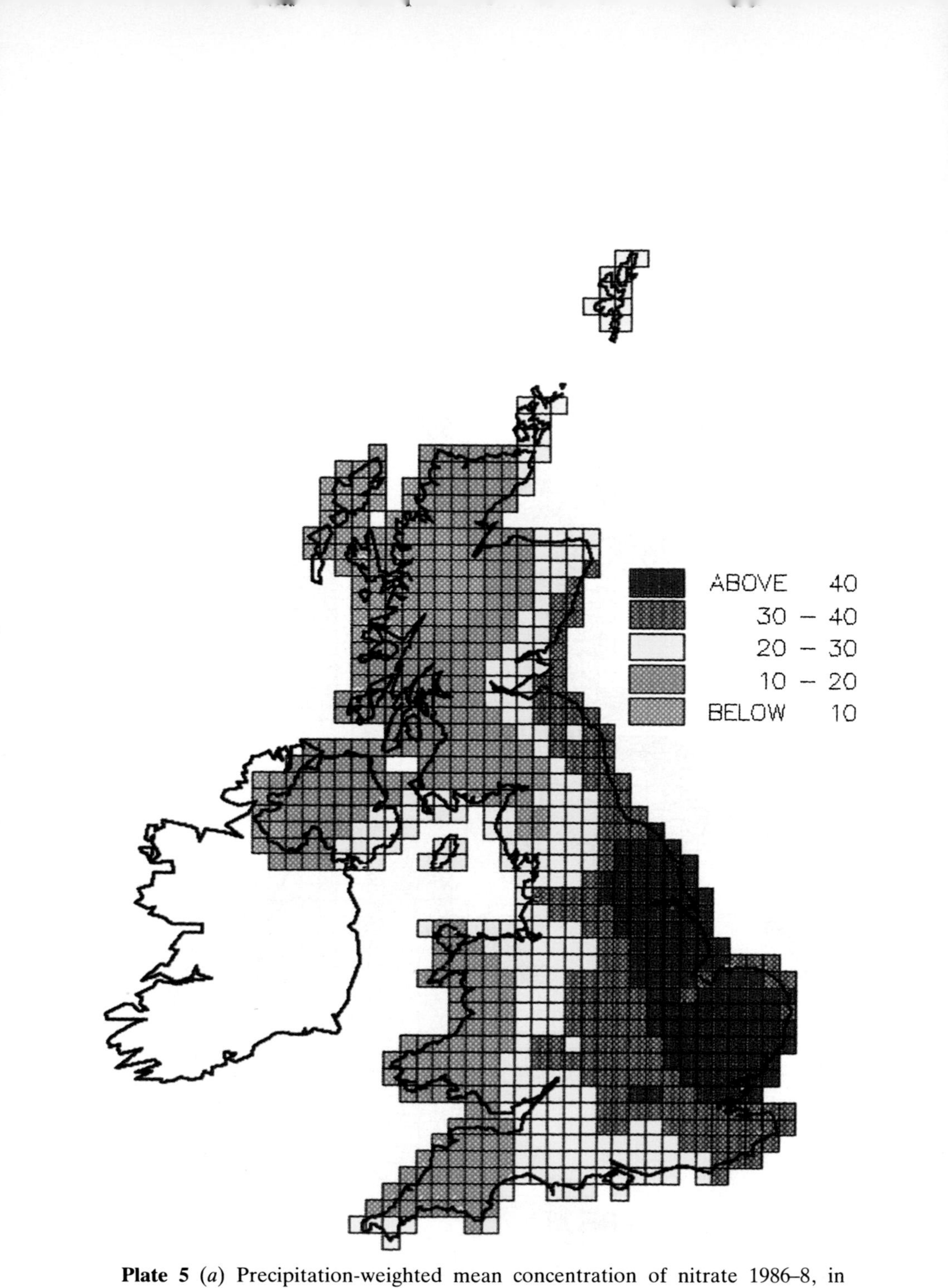

Plate 5 (*a*) Precipitation-weighted mean concentration of nitrate 1986–8, in microequivalents per litre.

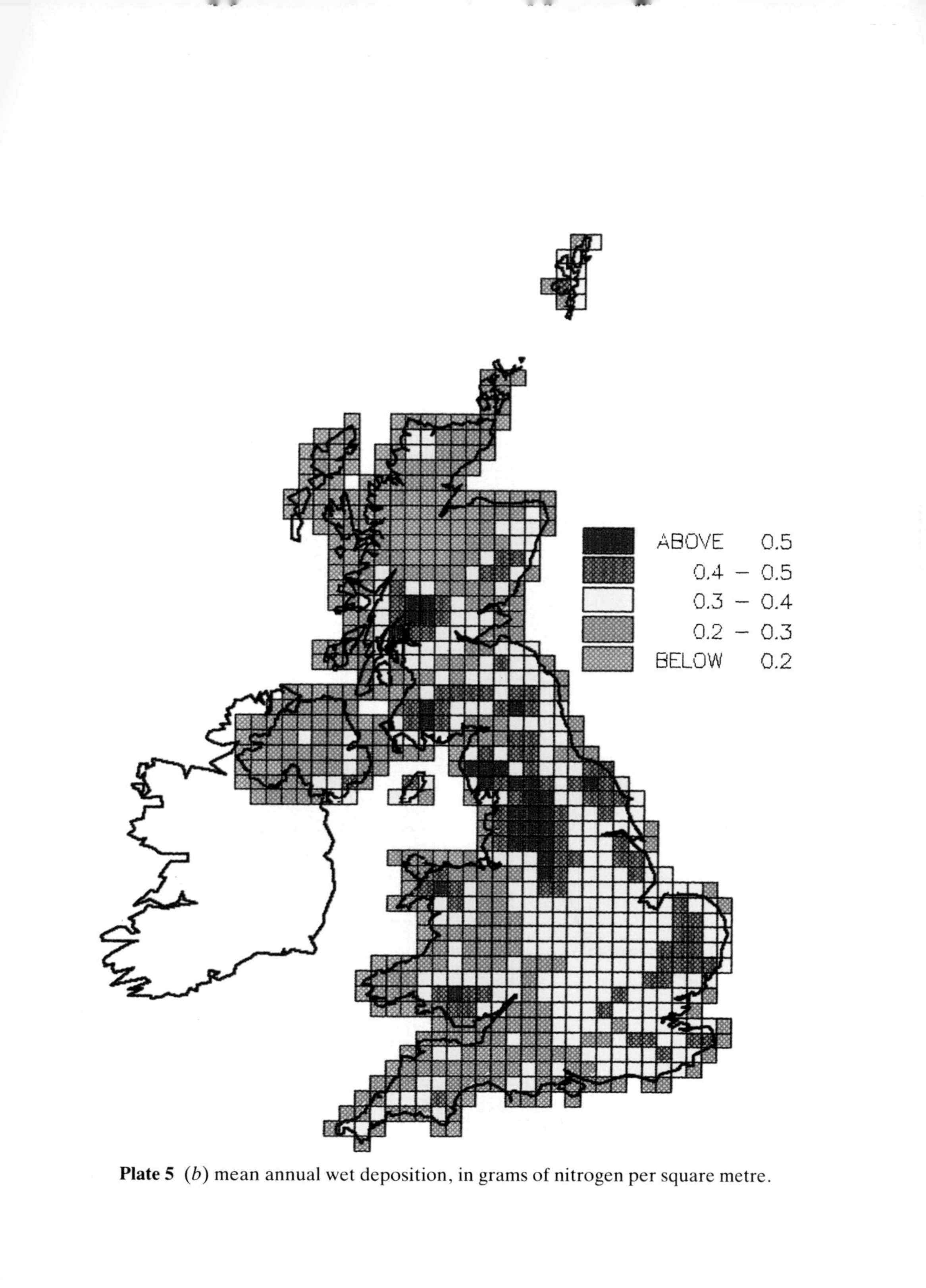

Plate 5 (*b*) mean annual wet deposition, in grams of nitrogen per square metre.

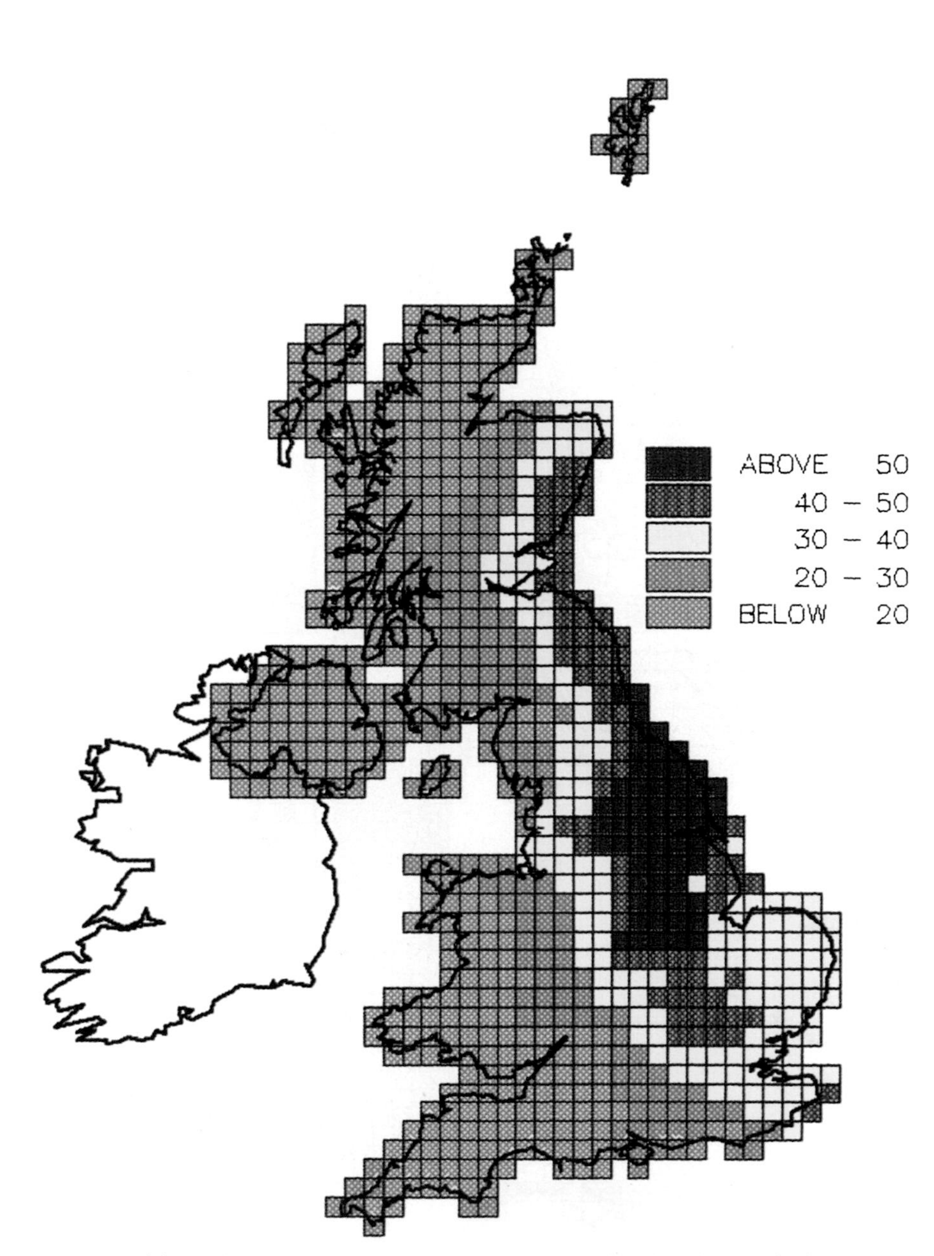

Plate 6 (*a*) Precipitation-weighted mean acidity, 1986–8, in microequivalents per litre.

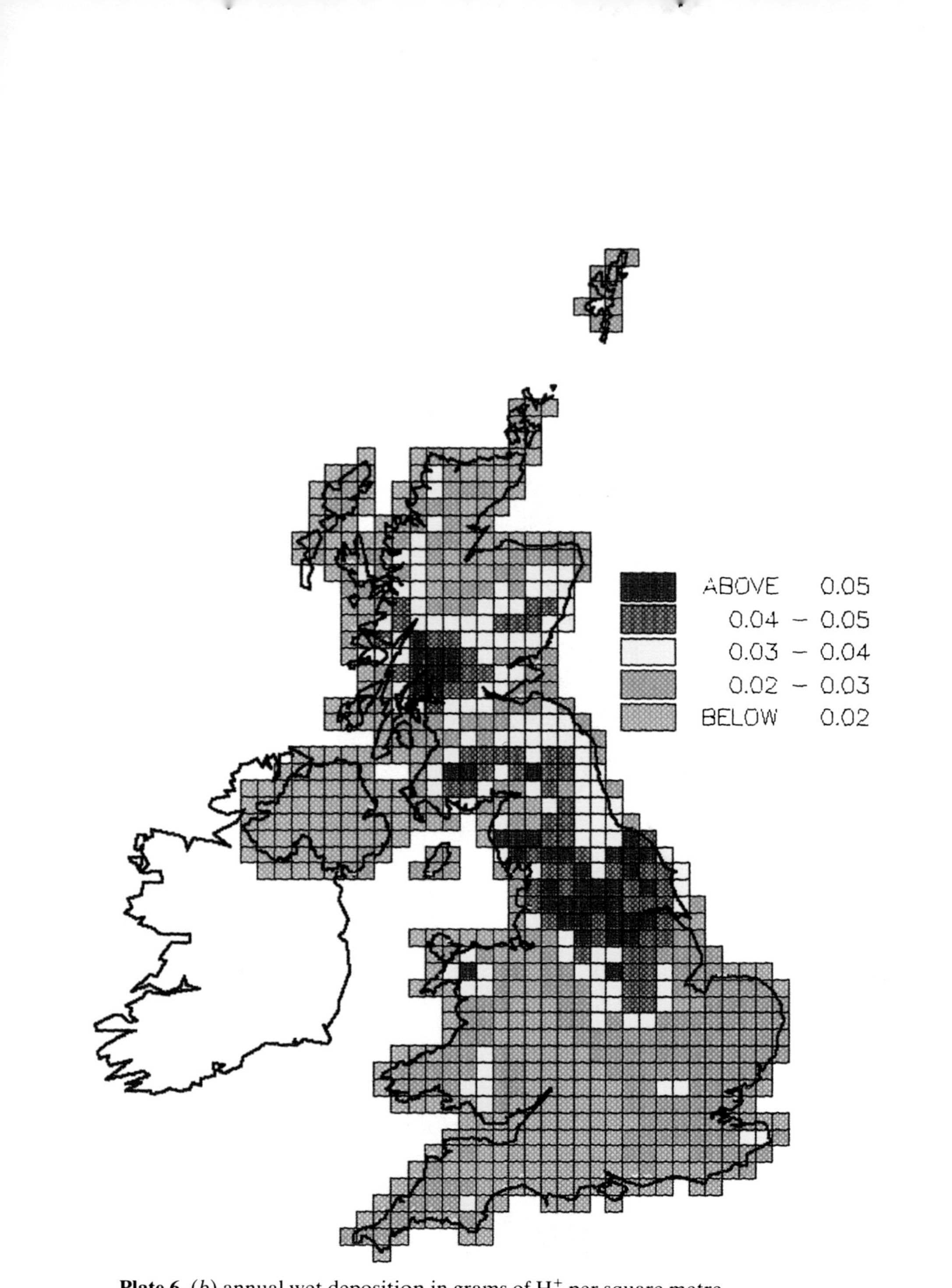

Plate 6 (*b*) annual wet deposition in grams of H^+ per square metre.

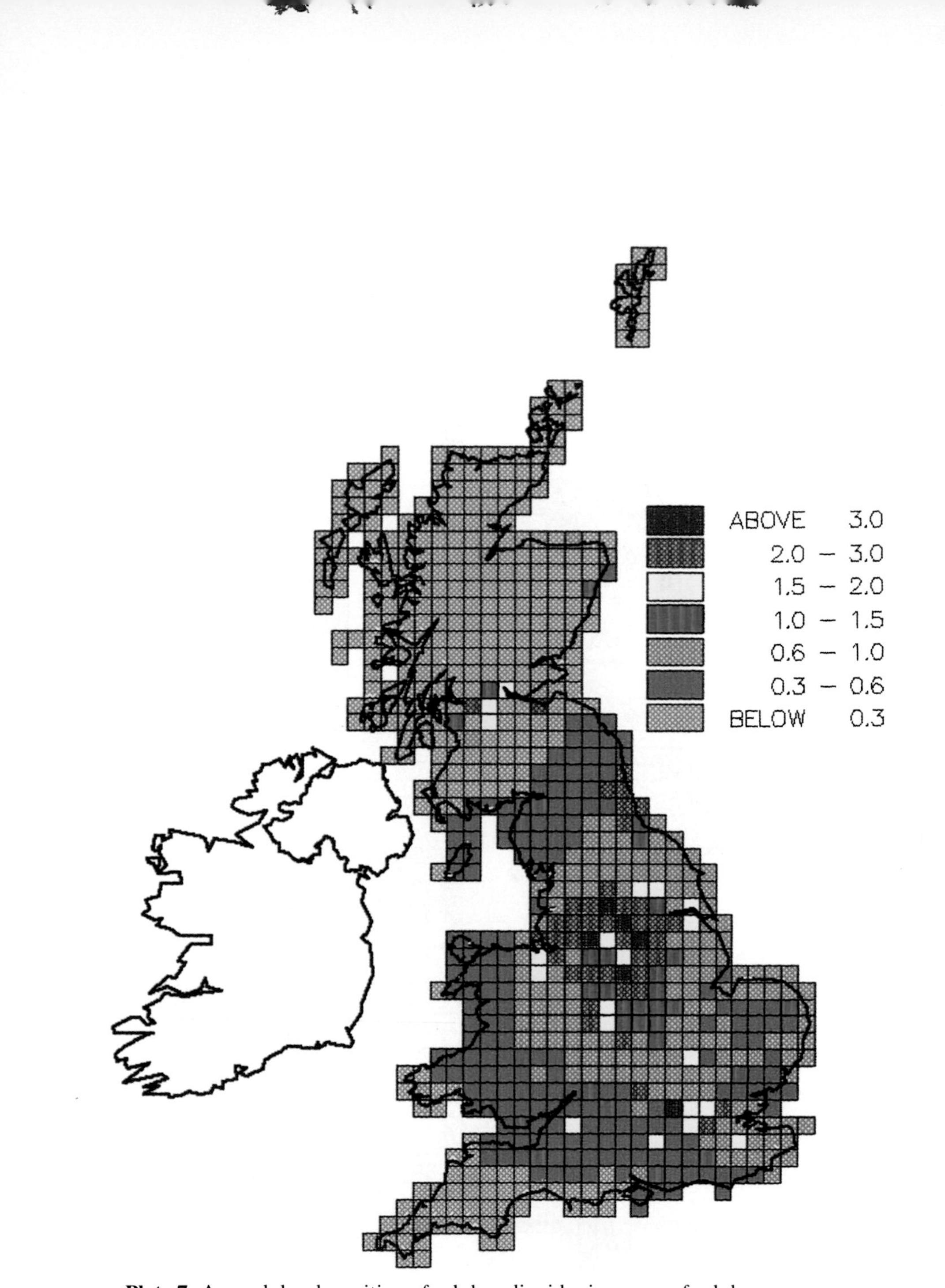

Plate 7 Annual dry deposition of sulphur dioxide, in grams of sulphur per square metre.

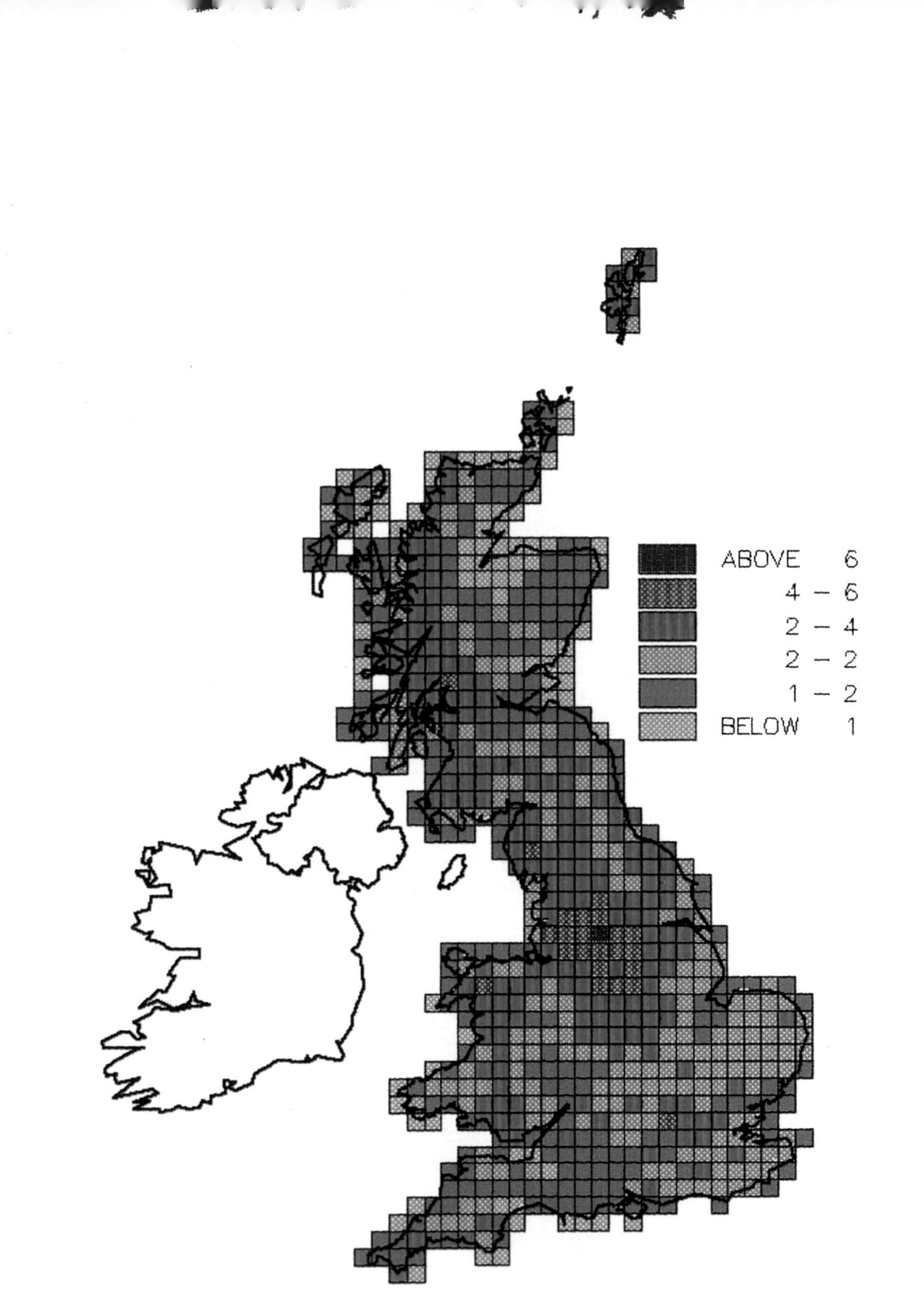

Plate 8 Total annual sulphur deposition, in grams of sulphur per square metre.

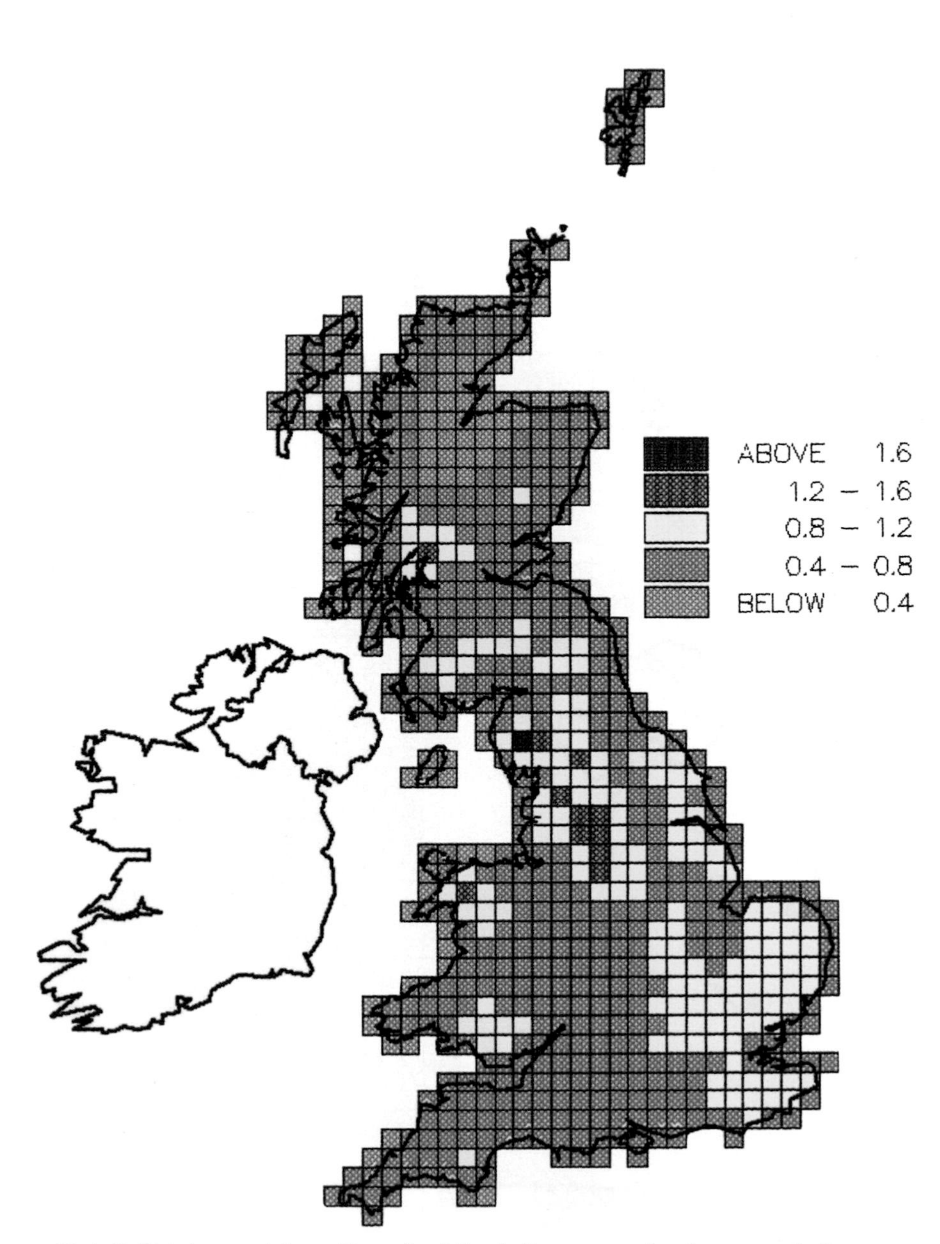

Plate 9 Total annual deposition of oxidized nitrogen species, in grams of nitrogen per square metre.

using the 1927 methods. On both occasions the upper horizons of older forests had a lower pH than those of younger forests, demonstrating the acidifying effect of forest growth. Moreover, the plotted line, showing the reduction in pH with tree age in 1984 is displaced downwards by 0.3–0.5 units compared with that for 1927. A similar reduction is found for the C horizon, which shows no discernible effect of tree age. These measurements offer fairly convincing, if not entirely unequivocal, evidence for the progressive acidification of soils due to acid deposition, which is high in this part of Sweden. Future studies of this kind should pay more attention to measured changes in base saturation, which are more reliable and probably a better indicator of long-term acidification than is soil pH.

5.4.7 ALUMINIUM CHEMISTRY IN ACID SOILS AND WATERS

The most important ecological effect of increasing the acidification of soils and surface waters is the mobilization of aluminium, which is toxic to aquatic life, especially fish, at concentrations of around 100 $\mu g\ l^{-1}$ that are often found in lakes and streams in Scandinavia, Scotland, and North America at pH values of around 5.

In the acid podzolic soils common in northern coniferous forests and formed from granitic bedrock, calcium and magnesium are in low concentration and the dominant exchangeable cation is aluminium. The relative abundance of the various aluminium species, inorganic and organic, depends on the pH, temperature, and the presence of inorganic and organic ligands. At low pH, below 5.0, H^+ ions from the soil water, when adsorbed on the surfaces of soil mineral particles such as gibbsite, $Al(OH)_3$, are exchanged for soluble Al^{3+} ions according to

$$Al(OH)_3 + 3H^+ \rightarrow Al^{3+} + 3H_2O.$$

When released into the water, the Al^{3+} cations capture OH^- ions to form hydroxyl — Al ions:

$$Al^{3+} \overset{OH^-}{\rightleftharpoons} AlOH^{2+} \overset{OH^-}{\rightleftharpoons} Al(OH)_2^+ \overset{OH^-}{\rightleftharpoons} Al(OH)_3,$$

the three cationic species existing in equal concentrations at pH 5 if in equilibrium with gibbsite, but Al^{3+} is dominant at lower pH. In deeper, less acid soil horizons, the aluminium tends to hydrolyse to form colloidal $Al(OH)_3$ and precipitate. The mobile aluminium cations may also capture F^- and SO_4^{2-} anions; in the latter case some of the aluminium may precipitate as insoluble basic aluminium sulphate $AlOHSO_4$.

Besides these inorganic, labile, water-soluble species, aluminium also appears in organically complexed, non-labile forms which are much less toxic to fish and other aquatic biota. The complexation of aluminium by organic acids, such as humic and fulvic acids, is poorly understood but, in very acid soils (pH < 4.5) it appears that complexation is much reduced and the toxic inorganic ions, especially Al^{3+} and $AlOH^{2+}$, predominate.

The total aluminium content of a soil solution can be determined by atomic absorption or by colorimetric analysis after acid digestion, or by neutron activation, which gives all the other trace metals as well. The organically complexed, monomeric aluminium can then be separated from the inorganic, monomeric forms by passage through an ion-exchange column where only the latter are retained. Then, having obtained the inorganic monomeric species by subtraction of the organic component from the total monomeric aluminium concentration, the concentrations of the individual species may be calculated on the assumption that they are all in solution equilibrium with the relevant solid phases.

However, a number of complications arise in applying this apparently straightforward technique, known as the Barnes–Driscoll method, to natural waters. The fractionation into monomeric inorganic and organic forms is often complicated by the presence of inorganic aluminium in colloidal, polymeric complexes at temperatures near 0 °C, by particles of microcrystalline gibbsite at higher temperatures, and also by alumina–organic complexes of high molecular weight. According to Lydersen and Salbu (1990), removal of these complexes by passing the solution through ultrafine, hollow-fibre filters leads to more consistent results in the determination of monomeric species. However, whilst there is experimental evidence to suggest that the high molecular weight species have no toxic effects on the yolk-sac larvae of salmon, colloidal aluminium may, at low temperatures, represent a sizeable fraction of the total and provide a potentially important reservoir of mobilizable aluminium, the importance of which in the biogeochemistry of aluminium cannot be safely ignored.

These complications apart, Tipping (1990) points out that the Barnes–Driscoll method requires that the organic complexes remain complexed as they pass through the ion-exchange column. He checks this by comparing exchange-column separation with equilibrium dialysis for the same solutions and finds that the exchange column tends to give lower values of organic aluminium, especially at high flow rates. He uses his results to correct those from exchange columns.

Because the method of calculating the concentrations of inorganic species assuming equilibrium with monomeric solid phases will not, in general, be valid in the presence of organic anionic complexes which tend to dominate in many natural waters, Tipping has developed models of aluminium–organic reactions, based on laboratory data, that provide quantitative descriptions of the binding of aluminium by aquatic humic substances (HS) as functions of their concentrations of pH. The models have been calibrated against field data and used to predict concentrations of organically complexed aluminium with an accuracy of ±20 per cent.

Table 5.3 Measured concentrations of aluminium species in streams at three sites in Norway

	pH	TOC (mg C l^{-1})	Al_t	Al_m	Al_o	Al_i ($\mu g\ l^{-1}$)
Høylandet	6.0	5.8(1.8)	106 (35)	57(22)	52(15)	6(6)
Birkenes	4.9	4.4(2.7)	410(291)	315(242)	101(65)	227(180)
Lomstjern	4.6	14 (5.9)	424(211)	320(162)	158(60)	158(112)

From Salbu *et al.* 1990. Values are given for both unfiltered and filtered samples, the latter in brackets. t total, m monomeric, o organic, i inorganic.

The results of these models can be incorporated into a scheme that allows the complete speciation of dissolved monomeric aluminium to be predicted in terms of Al^{3+} and its complexes with OH^-, F^-, and SO_4^{2-}, and humic substances.

But when all this is said and done, it is important to realize that, although equilibrium may be achieved between species in a stored sample, this state may not properly represent the conditions in a lake or river where the redistribution of species may not follow rapid changes in acidity. It therefore appears difficult to determine the concentrations of the various aluminium species in natural waters under realistic conditions that would be experienced by a fish.

Similar problems arise in soils and soil waters. Measurements on the release of aluminium as a function of pH from the four main soil horizons of the podzols of the acidified Birkenes catchment indicated that there were different rates of release and therefore different reservoirs of soluble aluminium in each of the horizons. Moreover, the results were not consistent with the assumption that the aluminium solubility was controlled by gibbsite, aluminosilicates or simple ion-exchange reactions. Better agreement was obtained when the activity was assumed to be regulated by organically complexed aluminium, but further investigations are needed.

Some measured concentrations of aluminium species at three sites in Norway, where the surface waters have very different values of pH and total organic carbon, are shown in Table 5.3. At Birkenes where the low pH is due mainly to the high input of mineral acids, the total concentration of aluminium is high, two thirds of the monomeric aluminium being inorganic at concentrations well above the toxic limit for fish. In the non-acid stream at Høylandet, the aluminium concentrations are low, very little being inorganic. In the acid waters of Lomstjern the total aluminium concentration is high and comparable with that at Birkenes, but with a higher proportion in organic form consistent with the higher concentrations of organic carbon.

5.5 References

Lydersen, E. and Salbu, B. (1990). In *The surface waters acidification programme*, (ed. B.J. Mason), pp. 245–50. Cambridge University Press.
Salbu, B., *et al.* (1990). In *The surface waters acidification programme*, (ed. B.J. Mason), pp. 251–4. Cambridge University Press.
Tipping, E. (1990). In *The surface waters acidification programme*, (ed. B.J. Mason), pp. 255–60. Cambridge University Press.

FURTHER READING

Aluminium chemistry

Driscoll, C.T. (1984). *International Journal of Environmental Analytical Chemistry*, **16**, 267–83.

Weathering studies

Bain, D.C., *et al.* (1990). In *The surface waters acidification programme*, (ed. B.J. Mason), pp. 223–36. Cambridge University Press.
J.I. Drever (ed.) (1985). *The chemistry of weathering*. Reidel, Dordrecht.
Jacks, G. In *The surface waters acidification programme*, (ed. B.J. Mason), pp. 215–22. Cambridge University Press.
Lundström, U. (1990). In *The surface waters acidification programme*, (ed. B.J. Mason), pp. 267–74. Cambridge University Press.

Soil–water interactions

Bache, B.W. (1984). *Philosophical Transactions of the Royal Society of London, Series B*, **305**, 393. 393–407.

6
Comparative catchment studies

6.1 The measurements and their interpretation

The total wet chemical deposition on a catchment comprises the product of the volume and chemical content of the rain and snow (bulk deposition) plus an enhancement contribution from the interception of cloud and fog water by trees, vegetation, etc. The dry deposition, in the form of particles and gases, is also enhanced by their capture by trees (especially conifers) and vegetation.

The inputs, in wet and dry deposition, measured both above and below the tree or vegetation canopy, of mineral acid, sulphate, nitrate, and base cations to the Scottish catchments and to the pristine Høylandet catchment are shown in Table 6.1, together with the concentrations of these ions in the bulk deposition, in the throughfall, in the leachate from the organic and mineral soil horizons, and in the stream water. The most relevant soil properties, i.e. pH, the reservoirs of base cations and exchangeable acidity, and concentrations of the various forms of extractable aluminium, are listed in Table 6.2. Similar, but more limited data for the other Scandinavian SWAP catchments are included.

The general and common features of these measurements in the various catchments, and also some of the marked differences between catchments, will now be summarized.

The input acid and sulphate deposition is enhanced by trees and vegetation at all the sites where the measurements were made, except at Høylandet. The high correlation between the concentrations of H^+ and non-marine SO_4^{2-} ions in rain water strongly suggests that the acidity is largely of anthropogenic origin. Although the volumes of rainfall are greater in the winter than in the summer in all catchments, higher concentrations of H^+, NH_4–N, and SO_4^{2-} are found in the summer owing, in part, to the more rapid oxidation of SO_2 by photochemically produced oxidants, and also to the greater release of ammonium during farming operations.

In passing through the tree canopies, the acidic rain water is partially neutralized in all catchments except Kelty, where uptake of NH_4^+ in exchange for H^+ by the Sitka spruce results in a drop in the pH of the throughfall. That this was not the case at Chon and Høylandet implies

Table 6.1 Chemistry of precipitation, throughfall, leachate and stream water for SWAP catchments

	pH	H^+	Ca^{2+}	Mg^{2+}	K^+	Total SO_4^{2-}	NO_3^-	Al_t ($\mu g\ l^{-1}$)	Al_l ($\mu g\ l^{-1}$)	TOC (mgl^{-1})
Allt a'Mharcaidh										
Input (meq $m^{-2}\ yr^{-1}$)										
above canopy		23	8		4	40				
below canopy		11			10	23				
Concentrations ($\mu eq\ l^{-1}$) in										
precipitation	4.64	23	10	16	4	37	13			
throughfall										
leachate from O	4.4		10–20							
leachate from BC	4.7		10–20							
stream water	6.5	0.3	40	26	6	52	1	30	4	2
L. Chon										
Input (meq $m^{-2}\ yr^{-1}$)										
above canopy		48			17	132				
below canopy		24			82	138				

Concentrations ($\mu eq\ l^{-1}$) in										
precipitation	4.54	29	11	21	4	53	19			
throughfall	4.8	15	50			92				
leachate from O	4.17	68			50	122				21 } DOC
leachate from BC	4.4	37			7	89				3 } DOC
stream water	5.08	8	72	44	5	93	10	129	47	5
Kelty										
Input ($meq\ m^{-2}\ yr^{-1}$)										
above canopy		72			14	140				
below canopy		80			47	230				
Concentrations ($\mu eq\ l^{-1}$) in										
precipitation	4.54	29	11	21	4	53	19			
throughfall	4.25	57	45			164				
leachate from O	4.0	100			9	115				12 } DOC
leachate from BC	4.1	80			9	73				13 } DOC
stream water	4.47	34	48	48	10	93	5	140	62	8

Table 6.1 (Cont.)

	pH	H^+	Ca^{2+}	Mg^{2+}	K^+	Total SO_4^{2-}	NO_3^-	Al_t ($\mu g\ l^{-1}$)	Al_l ($\mu g\ l^{-1}$)	TOC ($mg.l^{-1}$)
Høylandet										
Input (meq $m^{-2}\ yr^{-1}$)										
above canopy		20			6	29				
below canopy		4			26	29				
Concentrations ($\mu eq\ l^{-1}$) in										
precipitation	5.05	9	7	16		17	4			
throughfall	5.6	2.5				38				
leachate from O										
leachate from BC										
stream water	5.1	8	23	34	6	29	3	5–135	5–80	2–13
Birkenes I										
Input (meq $m^{-2}\ yr^{-1}$)						100				

Concentrations ($\mu eq\ l^{-1}$) in										
precipitation	4.28	52	9	13		61	38			
throughfall										
stream water	4.56	28	31	51	5	122	9	160–900	100–650	3–11
Svartberget										
Input ($meq\ m^{-2}\ yr^{-1}$)						36				
Concentrations ($\mu eq\ l^{-1}$) in										
precipitation	4.50	32	9	3		50	19			
throughfall										
stream water	4.7	20	108	57	8	114	1	25–300	15–50	5–30
Atna										
Input ($meq\ m^{-2}\ yr^{-1}$)						15				
Concentrations ($\mu eq\ l^{-1}$) in										
precipitation	4.7	20	2	1	1	30	11			
stream water	5.2	6	11	6	7	30		75	25	

Table 6.2 Soil chemical properties

	Soil horizon	pH	TBC[a] ($eq\ m^{-2}$)	Ex Acid	Ex Al	Extractable aluminium ($kg\ m^{-2}$)[a]			
						Acid soluble	Organic	Amorphous	Crystalline
Allt a'Mharcaidh									
Peaty podzol	O	4.20	0.56	7.55	0.28	0.77	2.11	1.29	0.60
	C	4.92	0.43						
Alpine podzol	H	4.10	0.55	3.97	0.65	0.43	2.39	0.41	0.74
	C	5.0	0.24	1.33	0.75				
Blanket peat	O1	3.86	3.60	22.8	0.38	0.09	0.24	0.01	0.11
	Egh	4.42	0.95	15.4					
L. Chon	LFH	4.23	2.0	9.86	0.16	0.41	2.31	0.23	0.86
	B/C	4.42	0.86	25.4	2.48				
Kelty	O1	3.70	0.87	34.4	4.2	0.22	0.92	0.35	0.17
	BC	4.61	1.49	16.8	—				
Høylandet	LF	3.94	3.08	8.7	0.13	1.11	2.48	0.90	0.46
	BC	5.08	0.63	43.9	2.25				

Based on Walker *et al.* 1990.
TBC, total base cations; Ex acid, exchangeable acidity; Ex Al, exchangeable aluminium; LFH, leaf fall humus.
[a] To a depth of 1 m.

that the dominant Norway spruce released basic cations, mainly K^+ rather than H^+, to balance the uptake of NH_4^+ ions.

The Høylandet, Birkenes, Allt a'Mharcaidh, Chon, and Kelty catchments are fairly representative of the acidic, upland podzolic soils of Scotland and Scandinavia, but differ by 1 to 5 in the average annual inputs of acidic deposition. The base cation content of these leached soils is low; only in the organic surface layers at Høylandet and Chon, and in the Allt a'Mharcaidh peat, does it exceed 100 meq kg^{-1}, and only in the peat does the total store exceed 10 eq m^{-2}. The exchangeable acidity, which increases in parallel with exchangeable aluminium in the B horizons, dominates and derives from the high organic content of the soils. Since the clay contents of all these soils are very low, less than 1 per cent in the mineral horizons, the H^+ and Al ions are bound largely to organic matter and sesquioxides, the organically complexed aluminium being 2.5–5.5 times as abundant as the acid-soluble aluminium. The fact that the latter decreases progressively in the sequence Høylandet, MPP, MAP, Chon, in increasing order for acid deposition, points strongly to leaching of the aluminium by the acidified soil water.

The base cation-exchange capacity of the soils is strongly correlated with the pH, with magnesium and calcium dominating in the ranking order Ca > Mg > K > Na, except at Kelty where Mg > Ca, probably because the foliage of the Sitka spruce efficiently captures marine salts.

Adsorption of sulphate by the soil removes the anion from solution and hence reduces the rate of cation leaching. Adsorption occurs mainly in the B horizons of the podzolic soils where the aluminium is concentrated and where sesquioxide surfaces may play an important role in the retention of sulphur by such reactions as

$$Al_2O_3 + 2SO_4^{2-} + 3H^+ \rightarrow 2AlOHSO_4 + OH^-,$$
$$Fe_2O_3 + 2SO_4^{2-} + 3H^+ \rightarrow 2FeSO_4 + 3OH^-.$$

It seems that the Mharcaidh catchment has the potential to adsorb large amounts of sulphate in the B horizons, but it also possesses large stores of readily leached aluminium.

The chemistry of the leachate from the organic and BC horizons of the forested Chon and Kelty catchments, summarized in Table 6.1, shows the organic horizons to be consistently the more acid owing to the contributions of organic acids. At Chon, the dissolved organic carbon (DOC) drops sharply as the water passes from the upper organic to the lower mineral horizons in this humus-iron podzol. The total output of DOC is greater from the peaty gleys of Kelty. The sulphate concentrations are greater in the organic horizons than in the BC flow at both sites, reflecting the mineralization of organic sulphur, but the flux of water and sulphate is very much less than in the BC horizon. However, the sulphur

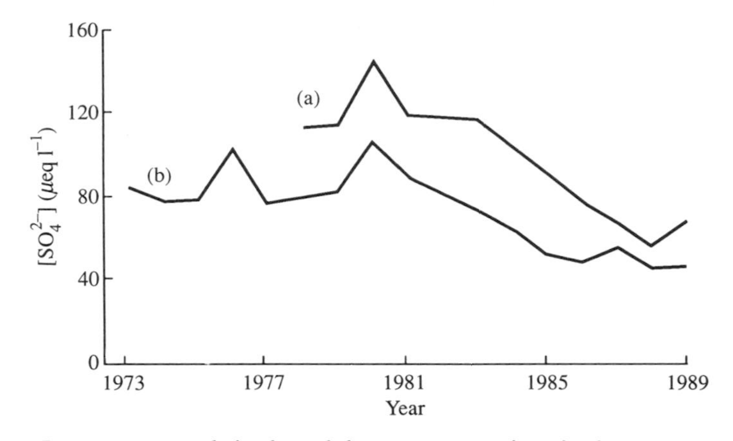

Fig. 6.1 Long-term trends in the sulphate concentrations in the stream water (a) and precipitation (b) at burn 2 in the Loch Ard catchment between 1973 and 1989. (From Harriman *et al.* 1990, p. 41, by permission of the Cambridge University Press.)

dynamics are different in the two catchments; although the inputs above the canopies and the concentrations in the streams are very similar, different processes are at work. The Sitka spruce at Kelty enhance the sulphate levels below the canopy, and the fact that the concentration in the throughfall exceeds that in the stream indicates that sulphate is being retained in the soil. At Chon the sulphate concentrations in the throughfall, leachate, and stream are nearly all equal, suggesting that the system is in balance, uptake by the roots being matched by release from the soils.

As to the long-term trends in the sulphate status of catchments, in Scotland the SO_4^{2-} concentrations in precipitation show a steep gradient from south-east to north-west, and an overall decline during the past decade in conformity with reduced emissions of SO_2. This change has been reflected in a reduction in loch and stream acidity at several sensitive acidified sites in south-west Scotland including Chon and Kelty. Figure 6.1 shows a parallel decline in SO_4^{2-} in both precipitation and streamwaters at an unafforested acidified site since 1980. During the last 5 years, stream and precipitation sulphate levels appear to have stabilized within the same time period, implying only a short delay between changes in input and stream response. However, two catchments, Chon and Birkenes, did show evidence of sulphate being released from soil reservoirs which would tend to delay the response of surface waters to reduced emissions of SO_2. It must be emphasized, however, that measurements of inputs and outputs for catchments have been made for only a few years and are subject to considerable error. Different methods of estimating these quantities can produce rather different results for the overall balance and different conclusions concerning retention or release of sulphate from the soils.

6.2 Main conclusions for each catchment

The main conclusions arising from the SWAP research programme for each site may be summarized as follows.

Høylandet. The acidic deposition is low and interception by the forest canopy results in partial neutralization of the rainwater. Acidification of the water as it percolates through the soil is limited by the relatively high base cation exchange capacity of the organic horizons. The stream water is further buffered by a high alkalinity base flow. Much of the soil acidity is of organic origin and much of the aluminium is organically complexed and so is non-toxic to fish.

Allt a'Mharcaidh. Acidic deposition is moderate and is enhanced in the upper parts of the catchment by the interception of cloud and fog water. The different vegetation communities receive different deposition loadings but there is a partial neutralization of the water as it passes through the vegetation canopies. It is not further neutralized as it passes through the soil profiles because much of it encounters peat, but stream water is buffered by the slow seepage of alkaline ground water into the base flow. The B soil horizons contain large stores of readily leached aluminium which help to retain some 25 per cent of the incoming sulphate.

L. Chon. The catchment is subject to high acidic deposition, the rain water is partially neutralized by the vegetation, is further neutralized as it percolates through the soil profile, and is strongly influenced by a deposit of calcium-rich dolerite in the upper reaches of the catchment. The overall result is stream water of pH 5 and $[Ca]/[Al_i] \approx 2$, which is acceptable to fish. Sulphate concentrations in throughfall, soil water and stream water are similar, indicating that root uptake and retention are balanced by release of sulphur from the soil.

Kelty. The large sulphate inputs and, to a lesser degree, the acidic deposition of this catchment are further increased under the Sitka spruce canopy. The fact that the sulphate concentration in the throughfall is double that in the emerging leachate implies substantial uptake by the trees and retention in the soil. But the lack of further buffering, such as occurs at L.Chon, results in the stream-water having a pH and $[Ca]/[Al_i]$ ratio that cannot support a fish population.

Birkenes. This site receives very high acidic deposition, the highest for all the Scandinavian SWAP catchments, and has the most acid stream water with pH averaging 4.56 but falling on occasions to 4.2. During high-flow episodes, concentrations of inorganic aluminium substantially exceed 100 $\mu g\ l^{-1}$ which, together with the high acidity, ensure that the stream is fishless. There has been a steady long-term decrease in the concentrations of Mg^{2+} and Ca^{2+}, but a decline in SO_4^{2-} concentrations in both precipitation and stream water has been observed only since 1985.

Atna. This is a very acid sensitive catchment in which lakes and streams can only just sustain fish populations. The input depositions of acid and sulphate are modest, about one-seventh of those at Birkenes. The stream water pH averages 5.2 but falls to as low as 4.7 in high-flow episodes following snow melt. The conductivity and concentrations of all ionic species are generally very low. Sulphate concentration in precipitation is the same as in stream water, allowing for an estimated 40 per cent evaporation, which implies some retention in the soil. Usually concentrations of aluminium and organic acids were low, but both increased during high-flow episodes when concentrations of Al_i exceeded 100 $\mu g\ l^{-1}$ and were probably deleterious to fish.

Svartberget. Here the deposition of acid and sulphate is higher than at Høylandet or Atna. However, the stream chemistry is dominated by natural organic acids and the aluminium is mainly bound to organic ligands. The sulphate input is approximately equal to the output, implying little retention in the soil.

In all the SWAP catchments, except Høylandet, the nitrate concentrations in the streams were much less than in the incoming precipitation, which suggests that most of the nitrate was taken up and retained by the vegetation.

6.3 References

Harriman, R., *et al.* (1990). In *The surface waters acidification programme*, (ed. B.J. Mason), pp. 31–45. Cambridge University Press.
Walker, T.A.B., *et al.* (1990). In *The surface waters acidification programme*, (ed. B.J. Mason), pp. 85–95. Cambridge University Press.

FURTHER READING

Bishop, K.H., *et al.* (1990). In *The surface waters acidification programme*, (ed. B.J. Mason), pp. 107–19. Cambridge University Press.
Christophersen, N., *et al.* (1990). In *The surface waters acidification programme*, (ed. B.J. Mason), pp. 97–106. Cambridge University Press.
Ferrier, R.C., *et al.* (1990). In *The surface waters acidification programme*, (ed. B.J. Mason), pp. 57–67. Cambridge University Press.

7
The history of lake acidification as reconstructed from palaeolimnological evidence

7.1 Introduction and overview

Palaeolimnologists are able to reconstruct the history of lake acidification by studying the remains of past acid-sensitive plant and animal communities in radioactively dated lake sediments laid down over centuries. Moreover, because the sediments also contain material derived from the atmosphere and the lake catchment, the changes in lake ecology can be correlated with changes in atmospheric deposition and in land use. A great advantage of this technique is that it can provide a continuous record stretching back over hundreds or even thousands of years.

Since the composition of a diatom community, containing several different taxa, is sensitive to the water chemistry, especially pH*, the history of lake pH can be inferred by studying the relative proportions of different diatoms in successive layers of sediment, calibrations being provided either by analyses of laboratory cultures grown under carefully controlled pH, or by correlating populations of living diatoms in the surface layers of lakes with contemporary measurements of pH and other relevant chemical species.

A cylindrical core of sediment is taken from a relatively undisturbed part of the lake, cut into 5 mm slices, and the diatoms classified and counted under the microscope. Each slice is dated using the lead isotope ^{210}Pb, which has a half-life of 22 years, well suited to dating sediments laid down since the industrial revolution. The fallout of ^{137}Cs from the atomic weapons tests in the 1950–60s also provides useful chronological markers, so that the mean age of a 5 mm slice laid down over about 5 years can usually be estimated to within ± 1 year. Carbon-14 dating can also be used as an additional check.

Although for the SWAP the emphasis has been on diatom analysis,

* The physiological basis for this sensitivity to pH is not fully understood. There is some evidence that the enhanced concentrations of aluminium that usually accompany low pH result in the precipitation of phosphorus and hence to phosphate starvation, at least for some diatom species.

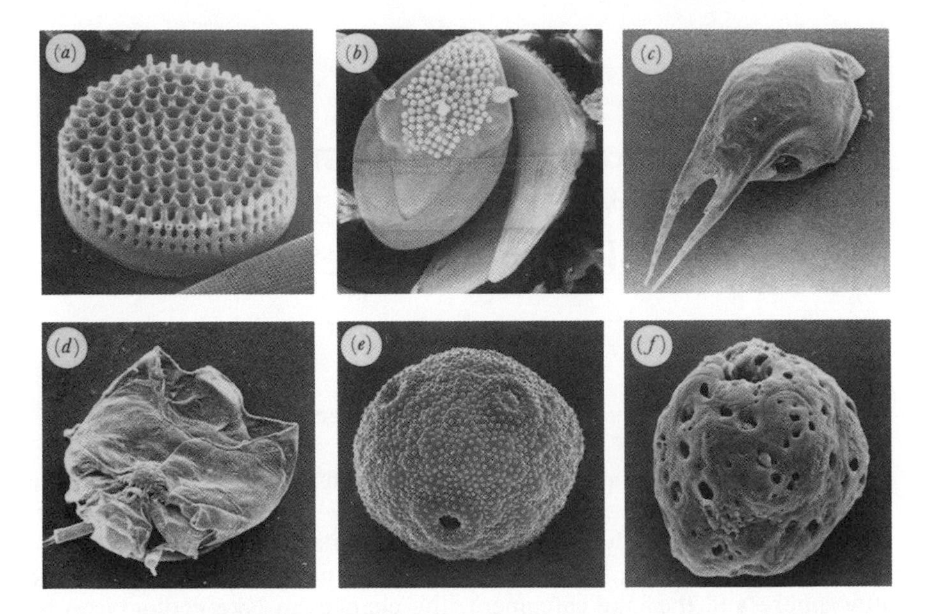

Fig. 7.1 Illustrations of microfossils and particles found in lake sediments used in reconstructing environmental changes: (a) diatom valve; (b) chrysophyte scale; (c) cladoceran head shield; (d) chironomid head capsule; (e) pollen grain; (f) spheroidal carbonaceous particle from coal combustion. (From Renberg and Battarbee 1990, p. 282, by permission of the Cambridge University Press.)

studies have been made of the remains of aquatic insects such as chironomids (midge larvae), and of single-cell flora covered by silica scales (chrysophytes), both of which are sensitive to pH, whilst the abundance of cladocerans (zooplankton) reflects past levels of fish predation, and counts of fish scales in the sediments indicate changes in fish populations, in particular when they die out.

Changes in the fluxes of atmospheric pollutants in the form of combustion products were indicated by analyses of trace metals, sulphur, polycyclic aromatic hydrocarbons (PAH), carbonaceous soot particles, and magnetic spherules. Changes in land use, vegetation, and forest cover were inferred from analyses of pollen grains. Some examples of microfossils and particles found in lake sediments and used in historical reconstructions of water chemistry and ecology are illustrated in Fig. 7.1.

The extensive SWAP studies, involving more than 40 scientists working on some 20 sites in Scotland, Norway, and Sweden (see Fig. 7.2), have established that many lakes in the three countries have undergone progressive acidification from *circa* 1850 until very recently. The magnitude of this acidification is appreciably greater than any that occurred during

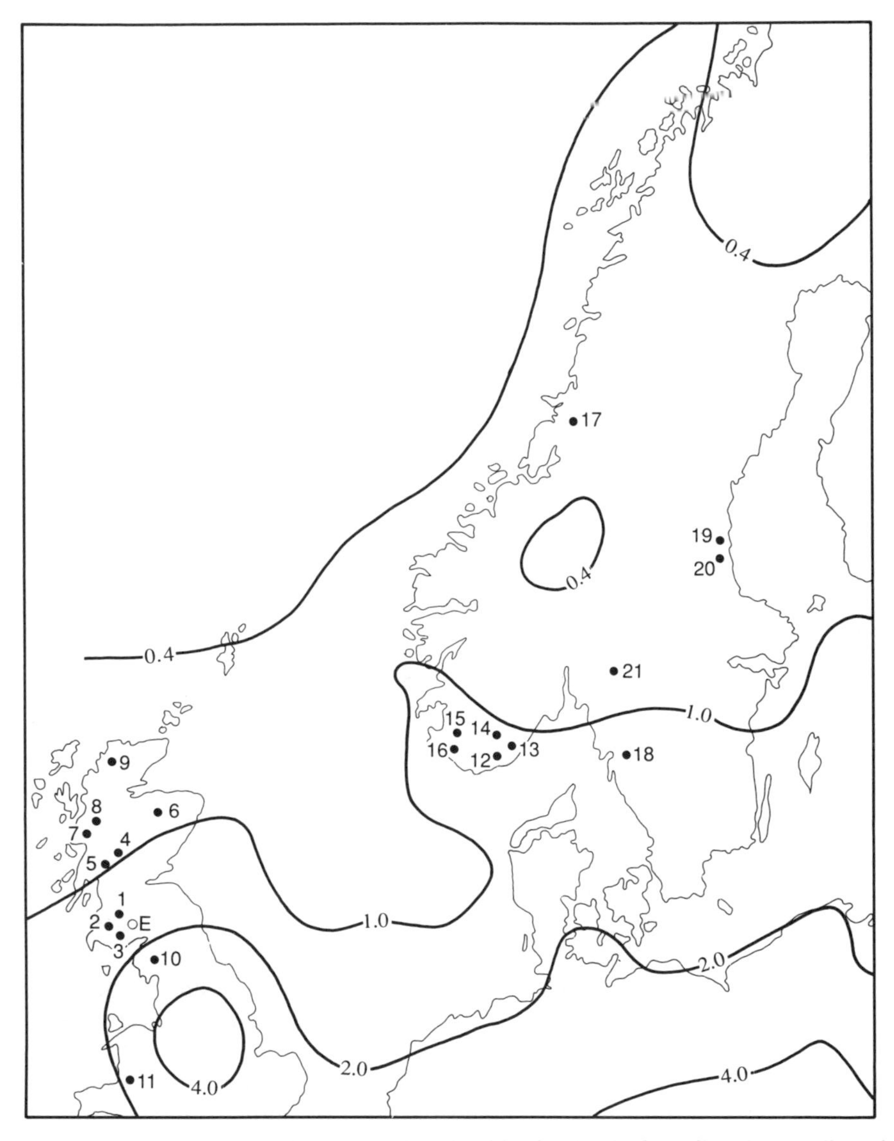

Fig. 7.2 The locations of the lakes involved in the Anglo-Scandinavian studies of sediment cores in relation to the isolines of sulphur deposition, in grams per square metre per year: 1, the Round Loch of Glenhead; 2, Loch Grannoch; 3, Loch Fleet; 4, Loch Tinker; 5, Loch Chon; 6, Loch Uaine; 7, Loch Doilet; 8, Lochan Dubh; 9, Loch Sionascaig; 10, Devoke Water; 11, Llyn Hir; 12, Vervatn; 13, Gulspettvatn; 14, Holmevatn; 15, 15, Holetjörn; 16, Ljosvatn; 17, Röyrtjörna; 18, Lilla Öresjön; 19, Sjösjön; 20, Lill Målsjön. (From Renberg and Battarbee 1990, p. 285, by permission of the Cambridge University Press.)

the previous 10 000 years and has marched in parallel with accelerated industrial development, as indicated by increases in several pollutants in the sediments. These changes and the extent of the inferred acidification are, in general, geographically correlated with the intensity of acid deposition and with the geochemical status of the catchment.

The overall evidence points strongly to acid deposition from the atmosphere being the main cause of lake and catchment acidification. There is little evidence to support the alternative hypothesis that acidification was generally brought about by changes in land use, such as afforestation, changes in agriculture, grazing, or burning of vegetation, although afforestation, especially by conifers, may accelerate acidification in areas of high acid deposition because of enhanced capture of acidic gases and aerosols by the forest canopy.

The evidence for these conclusions will now be reviewed.

7.2 The research strategy

The SWAP study sites are shown in Fig. 7.2 in relation to the contours of sulphur deposition. Integrated studies, involving most of the indicators described above, were undertaken at four sites in Scotland (Round Loch of Glenhead, Loch Tinker, Lochan Dubh, and Loch Uaine), two in Norway (Vervatn, Röyrtjorna), and Lilla Öresjön in Sweden, all of these being adjacent to stream catchments that were being studied from many other points of view.

Much effort was devoted to the improvement and standardization of techniques and protocols, especially in relation to diatom analysis. A large water chemistry and diatom calibration data set was constructed using data from more than 160 locations. Special data sets were used to train and cross-check the analysts, and workshops held to facilitate exchange and intercomparison of samples, intercalibration, and quality control, in order to ensure compatability and comparability of the results of the various research groups.

New statistical methods of pH reconstruction from changes in diatom communities, and of error estimation, were developed and applied to sediment cores from all SWAP sites. Detailed descriptions of these statistical methods are given by Birks *et al.* (1990) who estimate the standard errors in the pH determinations to range from 0.25 to 0.40 pH units.

Two projects in the UK and two in Sweden studied long-term acidification and its relationship to recent (post AD 1800) acidification. Five projects addressed the influence of catchment vegetation and land use on lake acidification. These included studies of afforestation in Scotland, spruce expansion in Sweden, grazing and burning in Norway and Scotland, and the acidification history of hill-top lakes with very small catchments in southern Norway. A further project considered the lake and sediment response to very recent (post 1970) changes, such as liming, forest fertilization and, most importantly, decreasing acid deposition resulting from lower emissions of SO_2.

These palaeolimnological investigations were led and coordinated in the UK by R.W. Battarbee, in Sweden by I. Renberg and in Norway by H.J.B. Birks. A full account of all this work was published by the Royal Society in 1990 in a special volume entitled, '*Palaeolimnology and lake acidification*'. (Battarbee *et al.* 1990). The main results are also summarized in '*The surface waters acidification programme*', published by Cambridge University Press in 1990 (Mason 1990). Individual references will therefore not be cited in this short review.

7.3 Long-term acidification

Analysis of diatom fossils in sediments has made it possible to trace the acidification history of some lakes over several thousand years. The results indicate that many of the currently acidic lakes were alkaline much earlier in their history while others have been acidic for much of their existence. In all the SWAP cases where acidification had taken place, the most rapid changes occurred soon after deglaciation owing both to the rapid leaching of cations from unweathered soils and the development of acidic organic soils as the vegetation changed from arctic tundra to forest. During the later stages of the post-glacial period, acidification proceeded much more slowly, often at less than 0.1 pH unit per millenium, suggesting a rough equilibrium between base cation production by weathering and cation loss by leaching. However, the pH of none of the lakes studied fell below 5.0 until after about 1850 when they became increasingly affected by acid deposition.

As an example, the long-term pH record for Lilla Öresjön in southwest Sweden since its formation 12 600 years ago is shown in Fig. 7.3. It shows four stages of acidification. It was alkaline for a long period from 12600–7800 BP, during which the pH fell slowly from 7.2 to 6.0. There followed a further period of slow natural acidification from 7800–2300 BP, during which the pH fell gradually from 6.0 to 5.2 with some short-term fluctuations. A period of higher pH intervened between 2300 BP and AD 1900 when it rose just above 6.0, coinciding with the appearance of cereal pollen in the sediments, together with other indicators of agricultural activity. The recent period of rapid acidification since 1900, and especially since 1950, during which pH fell to 4.5, will be discussed later.

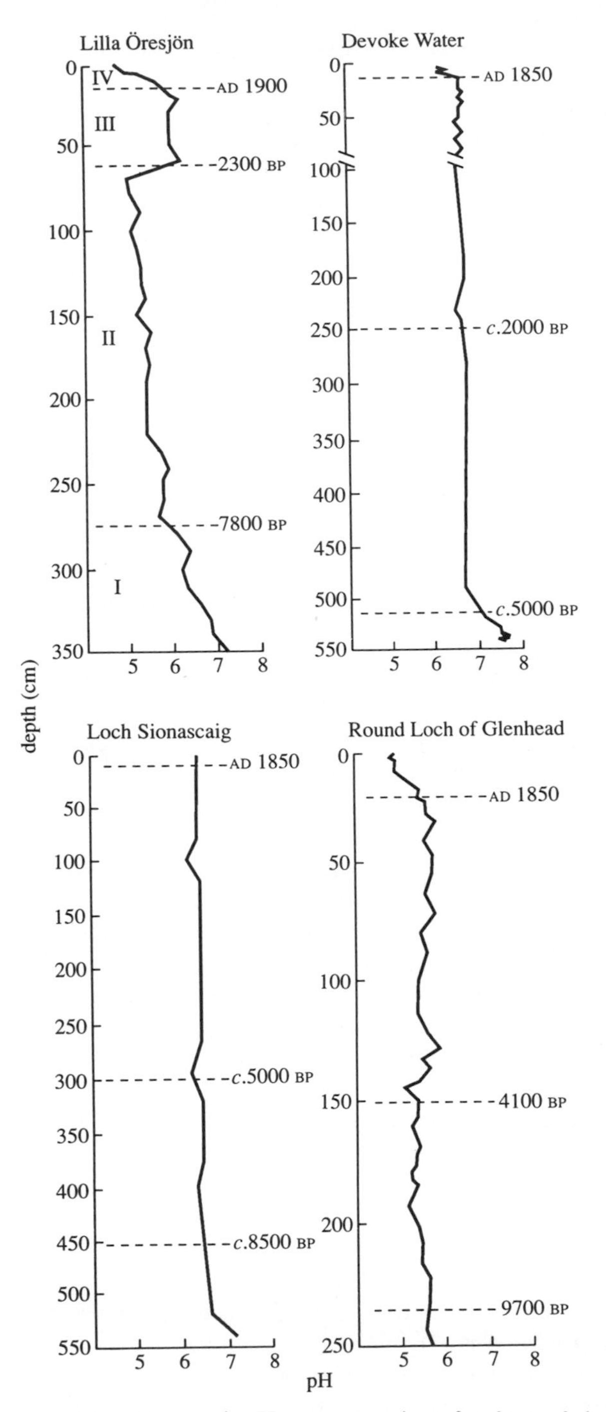

Fig. 7.3 Long-term (10 000 year) pH reconstructions for four of the lakes sited in Fig. 7.2. (From Renberg and Battarbee 1900, p. 290, by permission of the Cambridge University Press.)

7.4 Changes post AD 1800 at sites with high and low acid deposition

Studies carried out within the SWAP show that lakes in areas of low acid deposition have low concentrations of airborne pollutants in the sediments and have undergone no significant changes in pH. In contrast, lakes in areas of high acidic deposition have high concentrations of pollutants and have undergone severe acidification in recent times. In all cases there is a strong correlation between increasing trends in pollutants and decreasing trends in pH.

The pH profile for Round Loch of Glenhead in Fig. 7.4 shows a fall from 5.4 in 1870 to 5.0 in 1900 and to 4.8 in 1973, followed by a small increase in the last decade, reflecting perhaps the recent reduction in SO_2 emissions in the UK. These sharp falls in pH are accompanied by increases in the concentrations of zinc, lead, spheroidal carbonaceous particles, and magnetic minerals. Similar trends are shown in Fig. 7.5 for Loch Laidon on Rannoch Moor.

Figure 7.6 is instructive in that Loch Urr in Galloway receives high depositions of acid and industrial pollutants but the pH has not fallen below 6.5 since 1850 and is 6.8 today. This can be attributed to the fact that the loch rests on sedimentary rocks comparatively rich in calcium and magnesium which emphasizes the importance and role of the geochemistry of the catchment.

Biological changes associated with changes in lake acidity have been established for Lilla Öresjön, a lake in an area of high acidic deposition on the west coast of Sweden. The changes began gradually about AD 1900 when the pH was approximately 6.1, and a more acute phase followed in 1960 when the pH fell rapidly to its recent value of about 4.6. Roach and bream were present in the lake in the 1950s but died out in the 1960s. The diatom community and the scaled chrysophyte flora have changed markedly since 1900. Several species of cladoceran and chironomid communities have disappeared and have been replaced by new acid-tolerant species. Overall diversity has decreased. Similar results have been obtained from other acidified lakes in Scotland and southern Scandinavia.

Again there is close stratigraphic correspondence between these biological changes and indicators of deposition of atmospheric pollutants. The concentrations of metals and sulphur increased from the beginning of the nineteenth century to reach peak values during the 1960s and 1970s. Spheroidal carbonaceous particles and PAH, indicators of coal and oil combustion, and magnetic particles originating in fly ash, peaked between 1970 and 1980. Pollen analysis and other evidence point to considerable changes in vegetation having taken place around the lake, heather giving way to pine and spruce, so that coniferous forest covered 60 per cent of

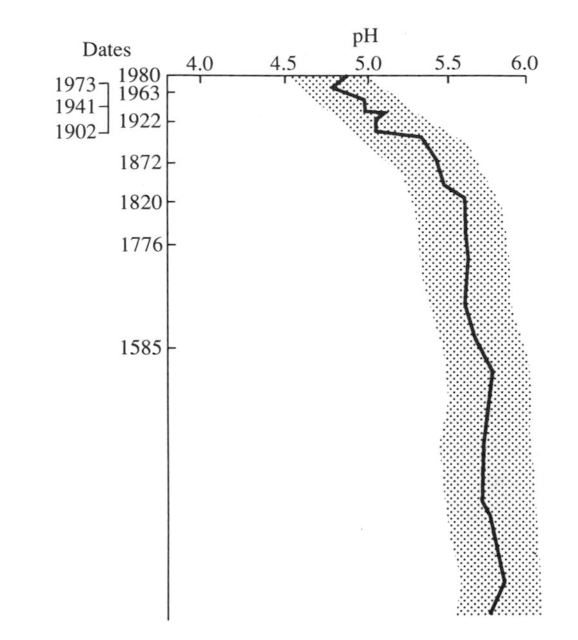

Fig. 7.4 pH reconstruction for the Round Loch of Glenhead, Galloway in more detail. (Courtesy of Dr R.W. Battarbee.)

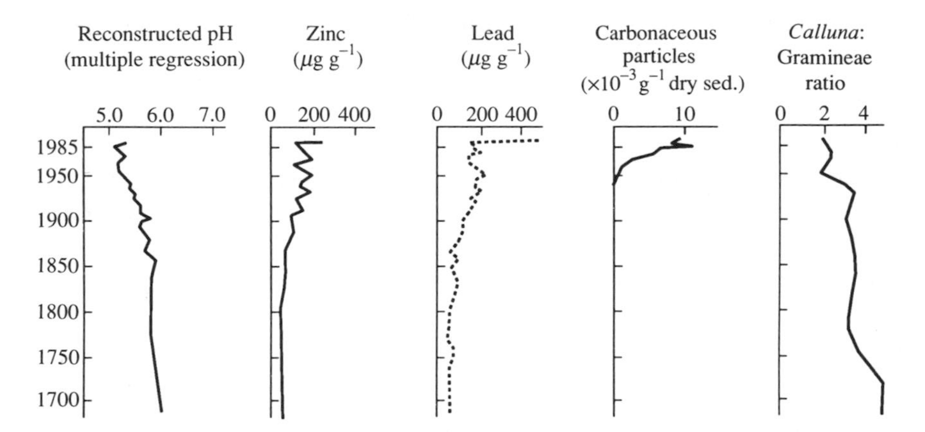

Fig. 7.5 Reconstructed histories of pH, lead, zinc, carbonaceous particles, and *Calluna*: Gramineae ratio for Loch Laidon. (Courtesy of Dr R.W. Battarbee.)

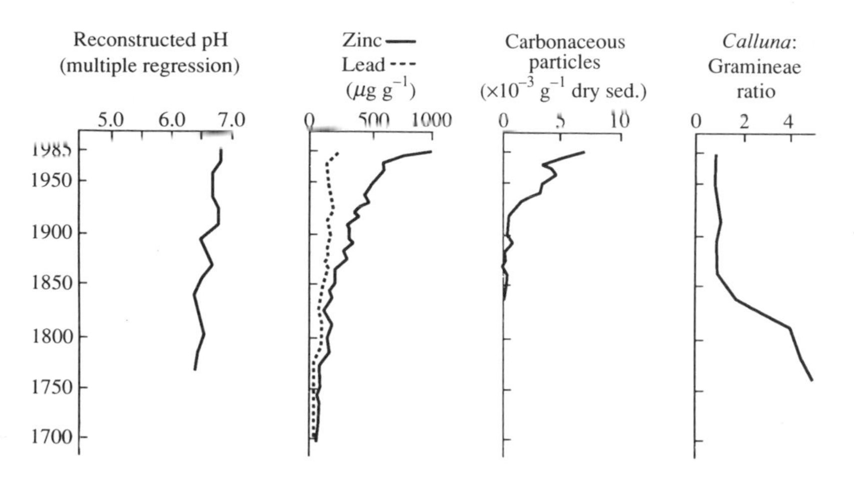

Fig. 7.6 Reconstructed histories of pH, lead, zinc, carbonaceous particles, and *Calluna:* Gramineae ratio for Loch Urr. (Courtesy of Dr R.W. Battarbee.)

the catchment in 1982. This may have led to some enhancement of acid deposition by the scavenging of atmospheric particles by the conifers. In contrast, the sediments of L. Röyrtjörna, which is close to the coast in central Norway and receives low acid deposition, show few changes in microfossil composition whilst diatom-inferred pH varies only between the relatively high values of 5.6 and 5.9. The cladoceran community also shows little change and the dominant species of chrysophytes are similar to those inhabiting Lilla Öresjön before acidification. The trace metal concentrations were very low and there was no significant change in sulphur deposition in the sediments over the last 100 years. Pollen analysis indicates that there has been little change in land use, vegetation or forest cover during this time.

7.5 The causes of lake acidification

Although there appears to be extensive and convincing evidence that the acidification, and especially recent acidification, of surface waters is largely due to man-made acidic deposition from the atmosphere, Rosenqvist (1978) has consistently argued against this hypothesis and attributes acidification of Norwegian lakes to a decline in the traditional agricultural practices of grazing and the burning of heathland. He argues that this has allowed natural vegetation to recover and cause acidification of the soil, both by taking up base cations from, and releasing H^+ ions into, the soil and soil water, and by increasing the accumulation of acid humus in the soil.

The SWAP investigations have found little evidence to support this hypothesis. Timberlid (1990) discovered that sheep grazing in south Norway, over the last 100 years, has decreased only below 600 m altitude. At higher levels, grazing has intensified but the acidification of lakes and streams has *increased*, in contradiction of the Rosenqvist hypothesis.

In the Cairngorms of Scotland there are several high corrie lakes with steep, boulder-strewn catchments, with little soil and vegetation, where no agriculture has ever taken place. Despite their isolation and lack of catchment changes, they are all acidified and their sediments contain high concentrations of atmospheric contaminants. Similar evidence comes from small lakes perched on hill-tops with very small catchment areas in Norway. Despite having experienced no catchment changes, they show clear signs of recent enhanced acidification with falls in pH from 5.0 to 4.5 between 1880 and 1940, with parallel increases in metallic and carbonaceous contaminants.

In order to assess whether land-use changes have played any significant role in lake acidification, one can study the influence of such changes in areas of, or during early periods of, very low acid deposition. Thus Jones *et al.* (1989) were able to show that, in the catchment of Round Loch of Glenhead, the formation of a blanket peatland between 3000 and 5000 years ago did not produce a change in pH as inferred from diatom analysis.

In the same vein, Renberg *et al.* (1990) selected eight Swedish lakes that are presently acidified with little peat in their catchments but large areas of spruce forest. Pollen analysis of the sediments established that natural immigration of spruce into this area started about 3000 years ago. Diatom analysis of the same cores found that expansion of the forest produced no shift towards more acid-resistant taxa and hence no change in pH of the lake waters. The inference is that, in the absence of acid deposition, afforestation did not cause lake acidification.

Comparisons between adjacent afforested and moorland streams in the uplands of Scotland have shown that the former have lower pH, higher aluminium and sulphate concentrations, and poorer fish populations than moorland streams. These differences have been mainly attributed either to the direct effect of forest growth or to the indirect effect of the forest canopy scavenging acidic gases and aerosols from the atmosphere, or a combination of both. In an attempt to separate these factors in time and space, a palaeolimnological study was devised in the SWAP to compare the acidification histories of afforested L. Chon and non-afforested L. Tinker in the Trossachs area of high acid deposition, with the afforested L. Doilet and non-afforested Lochan Dubh in north-west Scotland where the acid deposition is much lower. The pH reconstructions for the four sites are shown in Fig. 7.7. All four show evidence of acidification over the past century and increased concentrations of trace metals and carbon particles although the latter are very low in the north-western sites.

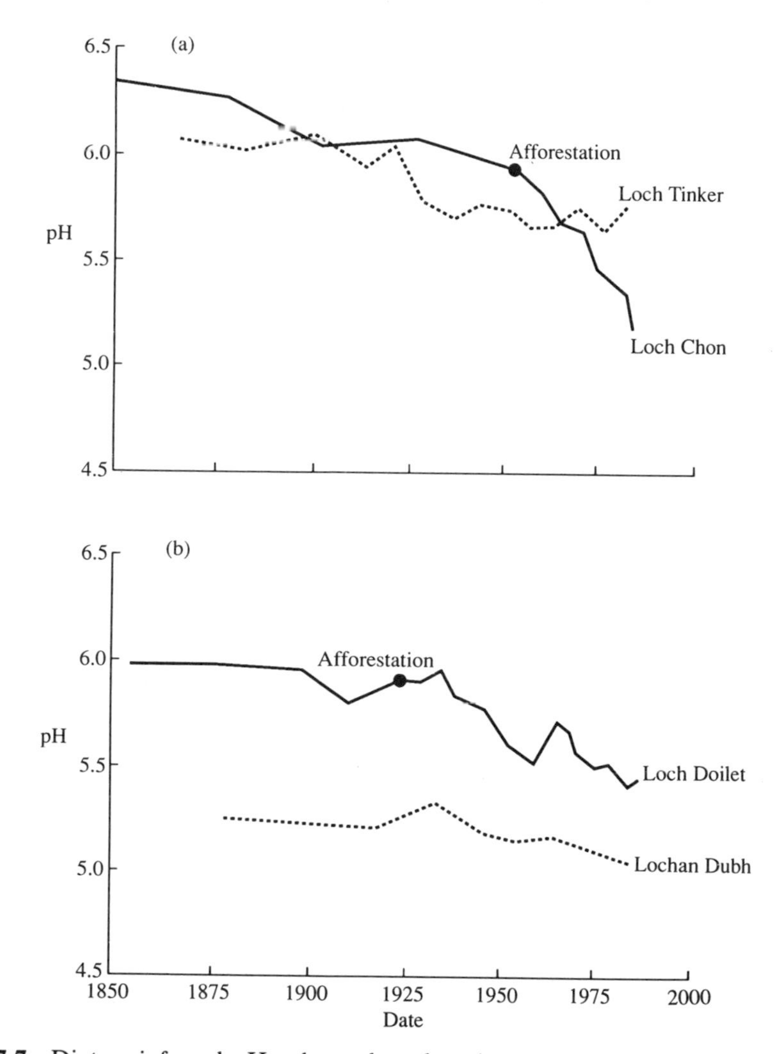

Fig. 7.7 Diatom-inferred pH values plotted against time for (a) two high acid deposition sites, Loch Chon (afforested 1950s) and Loch Tinker (non-afforested), and (b) two low acid deposition sites, Loch Doilet (afforested 1920s) and Lochan Dubh (non-afforested). (From Kreiser *et al.* 1990, p. 156, by permission of The Royal Society, London.)

In the high sulphur deposition region, acidification of L. Chon was greatly accelerated after afforestation whilst pH at the moorland 'control' site remained almost unchanged. Since both sites have similar base cation chemistry, and showed similar diatom-inferred pH trends before afforestation, the inference is that the recent enhanced acidification of L. Chon was due to afforestation.

In the region of low sulphur deposition, acidification at Loch Doilet occurred only after afforestation, but the fall in pH is not nearly so large compared with the control site as for L. Chon. For these reasons, and because L. Doilet is more sensitive to acidification than L. Chon, having a $[Ca^{2+}]$ of only 40 μeq l^{-1} compared with L. Chon's 80 μeq l^{-1}, the conclusion is that afforestation alone does not produce severe acidification but only when combined with high acidic deposition, which the forest canopy enhances by scavenging.

In conclusion, Renberg and Battarbee, in summarizing and assessing the results of all the SWAP palaeolimnological studies state that, apart from the case of the conifer afforestation just described, all tests designed to disprove the land-use hypothesis have succeeded, and all those designed to disprove the acid deposition hypothesis have failed. Charles (1990) comes to a similar conclusion from studies on sediment cores from more than 100 lakes in North America, although he concedes that catchment changes and natural long-term processes may play a minor role.

At individual sites recent acidification always post-dates major industrialization in the late eighteenth and early nineteenth centuries, the diatom response always occurs after the first signs of atmospheric contamination in the sediment record, and strongly acidified sites are always strongly contaminated by trace metals and carbon particles. Moreover, recently acidified sites are found in areas of high, and never in areas of low, sulphur deposition. Also Battarbee finds a dose–response relationship between sulphur deposition and lake acidification for UK sites if lake sensitivity is taken into account. This is illustrated in Fig. 7.8 which indicates that recent acidification has occurred on non-afforested sites only if the ratio of Ca^{2+} concentration in μeq l^{-1} to sulphur deposition in g m^{-2} yr^{-1} is less than 70.

7.6 Evidence of recent recovery in lake-water chemistry

Since the early 1970s, UK emissions of sulphur have declined by about 30 per cent and this has led to reduced acidity and sulphate concentrations in some Scottish lakes. It is instructive to examine and compare the responses of two acidified lakes in south-west Scotland, Round Loch of Glenhead located in undisturbed peatland where reduced sulphur deposition has been the only significant change during the past two decades, and nearby Loch Grannoch, where additional changes have resulted from extensive afforestation in the 1960s and 1970s.

At Round Loch of Glenhead, pH of the precipitation and lake water have both increased by between 0.2 and 0.3 units since the 1970s and sulphate has decreased by 40 per cent and 33 per cent respectively. During the summer of 1988, the pH exceeded 5.0 for the first time since

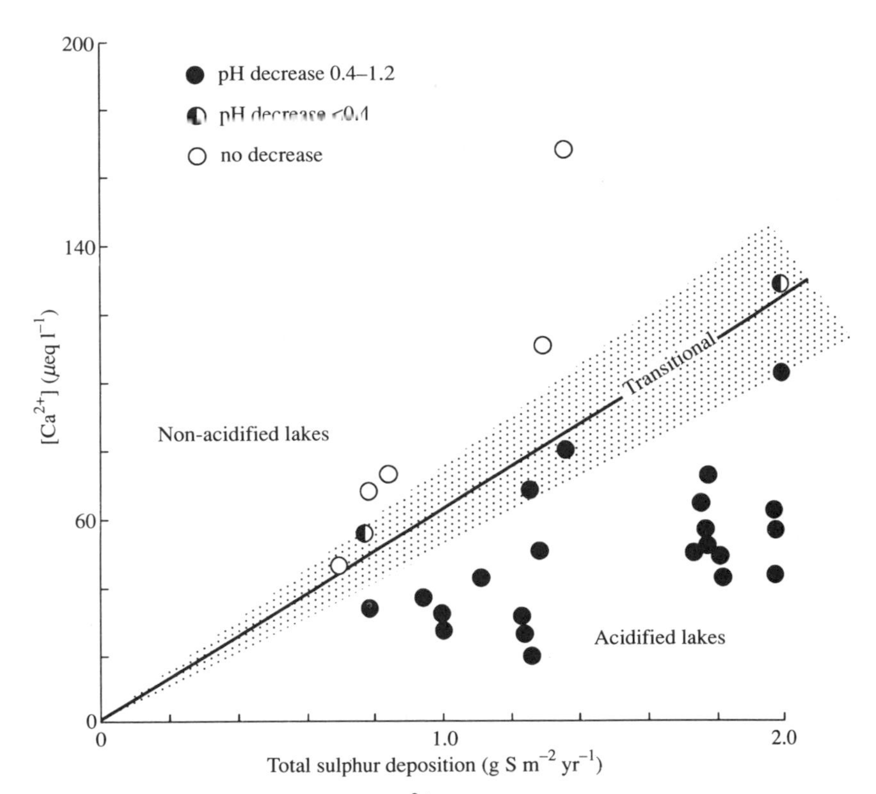

Fig. 7.8 The relation between Ca^{2+} concentrations in lake waters and total deposition of sulphur for several sites in the UK: ○, non-acidified lakes; ◐, slightly acidified lakes; ●, acidified lakes. (From Renberg and Battarbee 1990, p. 296, by permission of Cambridge University Press.)

measurements began. However, no marked changes appeared in the post 1970 layers of the sediment cores, probably because most of the pH increase occurred in winter rather than in the diatom growing season. The concentration of carbonaceous particles increased strongly after 1900 to reach peak values in the 1960s before declining after 1970.

At Loch Grannoch, the annual mean pH of the lake water has increased by about 0.2 units and the sulphate concentration has declined about 30 per cent since 1978. The pH, as inferred from diatom analysis, fell from 5.3 in 1974 to 4.6 in 1984 but has changed little in the last decade. The fact that non-marine sulphate concentration is consistently higher, the pH changes less marked and more variable than at Round Loch of Glenhead, may probably be attributed to the effects of afforestation which tends to enhance acid deposition. However, unlike Round Loch of Glenhead, a major change in the diatom community began in the late 1970s and culminated in 1988 with a strong incursion of diatoms indicative of peatland disturbance and nutrient enrichment due to the application of phosphorus and potassium fertilizers to the newly planted forest.

Evidence of recent changes in lake-water chemistry comes also from direct chemical measurements. Forsberg and Morling (1989) found that, during 1967–77, sulphate concentrations in lakes in western Sweden increased almost linearly with time from 100 to 300 $\mu eq\ l^{-1}$ and the pH fell from 6.8 to 4.5 in L. Oxsjön. After 1977, $[SO_4]$ decreased linearly to 200 $\mu eq\ l^{-1}$ and the pH rose by 0.3–0.4 units as the result of reduced wet acid deposition. This was the first evidence for reversal of lake acidification in Sweden.

The pH and alkalinity of about 700 lakes in south Norway were measured in 1974. A new survey of those lakes and some additional ones in a total of some 1000 lakes was made by Henriksen *et al.* (1990) in 1986, in order to determine any changes that have occurred during the intervening 12 years during which there has been a marked reduction in SO_2 emissions in western Europe. The water samples were analysed for all the main chemical species, total organic carbon (TOC), inorganic and organic aluminium. Of the 1000 lakes, 400 had $pH < 5.0$ and 600 $pH < 5.5$. Half had $[SO_4] > 50\ \mu eq\ l^{-1}$.

The new survey shows that the pH of lakes in south Norway has *not* changed significantly since 1974. The sulphate concentration is markedly lower in the most highly polluted lakes in the southernmost coastal region, which receives the highest SO_4 deposition. However, NO_3 concentrations have almost doubled.

In lakes further from the coast there has been little change in either $[SO_4]$ or $[NO_3]$. The increase in $[NO_3]$ is not due to increased nitrate in the rainwater, but probably due to the catchments with sparse vegetation and thin soils having become saturated with nitrate that is normally taken up by vegetation and micro-organisms in the soil. The excess NO_3^- ions probably act as mobile ions like SO_4^{2-} and contribute to acidification in the same way. The highly acidic lakes also have high concentrations of inorganic aluminium that are toxic to fish.

7.7 References

R.W. Battarbee, B.J. Mason, I. Renberg, and J.F. Talling (eds.) (1990). Palaeolimnology and lake acidification. The Royal Society, London.

Birks, H.J.B., *et al.* (1990). In *The surface waters acidification programme*, (ed. B.J. Mason), pp. 301–13. Cambridge University Press.

Charles, D.F. (1990). In *Palaeolimnology and lake acidification*, (ed. R.W. Battarbee, B.J. Mason, I. Renberg, and J.F. Talling), pp. 177–87. The Royal Society, London.

Henriksen A., *et al.* (1990). In *The surface waters acidification programme*, (ed. B.J. Mason), pp. 119–212. Cambridge University Press.
Jones, V.J., Stevenson, A.C., and Battarbee, R.W. (1989). *J. Ecology*, **77**, 1–23.
Kreiser, A.M., *et al.* (1990). In Palaeolimnology and lake acidification, (ed. R.W. Battarbee, B.J. Mason, I. Renberg, and J.F. Talling), pp. 151–6. The Royal Society, London.
Mason, B.J. (ed.) (1990). *The surface waters acidification programme*. Cambridge University Press.
Renberg, I., *et al.* (1990). In *Palaeolimnology and lake acidification*, (ed. R.W. Batterbee, B.J. Mason, I. Renberg, and J.F. Talling), pp. 145–6, The Royal Society, London.
Renberg, I. and Battarbee, R.W. (1990). In *The surface waters acidification programme*, (ed. B.J. Mason), pp. 281–300. Cambridge University Press. (Contains copious references).
Rosenqvist, I. Th. (1978). *Science of the Total Environment*, **10**, 39–49.
Timberlid, J.A., *et al.* (1990). In *Palaeolimnology and lake acidification*, (ed. R.W. Battarbee, B.J. Mason, I. Renberg, and J.F. Talling), pp. 137–41. The Royal Society, London.

8
The effects of acidification on fish and other aquatic life

8.1 Fishery status of lakes and streams in relation to water chemistry

The fishery status of a lake or stream can be assessed in terms of the presence or absence of various species, the number density, size (mass) and age distribution of current populations, and their rates of hatching, recruitment, growth, and death. These measures can be related to the water quality in terms of the concentrations of the major ionic species (especially H^+, Ca^{2+}, Mg^{2+}, Al species, SO_4^{2-}, NO_3^-), and the acidity or alkalinity.

A number of such studies has been made in recent decades in several countries, notably in Norway, UK, USA, and Canada, but usually only intermittently and over rather short periods, so that there are few records of adequate continuity and duration from which to draw confident conclusions on the actual causes of decline in fish populations. However, some common features are beginning to emerge that point to most likely causes, and these are reinforced by the results of controlled laboratory experiments. The situation is complicated by the recent discovery that major fish kills tend to be associated with short episodes of high acidity, often accompanied by high concentrations of aluminium and low concentrations of calcium (see Chapter 4 and Section 8.2 below).

A review of field studies of the fishery status of acid waters in the UK has been published recently by Turnpenny (1989). Although there are rather few instances in the UK where loss or decline of fish populations can be confidently ascribed to acidification, the results of a survey of 22 upland lochs and associated streams in south-west Scotland indicated that such a decline had occurred in a number of lochs overlying granitic rocks over recent decades and was accompanied by a decline in pH of 0.5–1.2 units, as inferred from changes in diatom communities in lake sediments (see Chapter 7).

Records of Loch Fleet show a progressive decline in the catch of brown trout since about 1950 and no fish caught after 1972. Survival studies of artificially implanted brown trout eggs and fry in 1984–5 demonstrated

Table 8.1 UK upland streams: chemistry of five stream groups categorized by pH (the values are means, with ranges in parentheses)

	Group 1 pH < 5.0	Group 2 pH 5.01–5.50	Group 3 pH 5.51–6.00	Group 4 pH 6.01–6.5	Group 5 pH > 6.50
Number of sites in sample	25	36	30	48	42
Mean pH	4.63 (4.16–5.00)	5.26 (5.02–5.50)	5.72 (5.53–6.0)	6.23 (6.01–6.50)	6.96 (6.54–7.56)
Minimum pH	4.21 (3.70–4.70)	4.61 (4.00–5.00)	4.78 (3.70–5.70)	5.51 (3.82–6.00)	6.50 (5.80–7.18)
Calcium (mg l^{-1})	1.5 (0.74–3.2)	2.2 (0.80–6.4)	2.6 (1.0–6.2)	3.6 (0.90–14)	7.9 (2.1–27)
Total aluminium ($\mu g\ l^{-1}$)	231 (81–550)	137 (80–280)	131 (17–530)	90 (13–380)	79 (10–300)

(After Turnpenny 1989.)

that the loch and its main spawning stream were acutely toxic to these early life stages owing to low pH (4.0–4.5), low calcium (0.5–2.0 mg l^{-1}) and high aluminium (around 200 $\mu g\ l^{-1}$). There is also evidence of a widespread decline in salmon fisheries in Scotland, especially in areas of heavy coniferous afforestation, which are known to enhance the acidity of precipitation and drainage waters.

Similar results have been observed in lake and river fisheries of central and west Wales, where the bulk of the observations have been made in upland streams because these are the main spawning sites for salmonids and are subject to the greatest extremes of acidity. Table 8.1 presents data on the water chemistry of upland streams in Wales and the Pennines, classified into five groups according to mean pH. The overall pattern is one of decreasing concentrations of calcium and increasing concentrations of total aluminium with decreasing values of pH. The fishery status of these streams is shown in Fig. 8.1, the percentage of fishless streams decreasing, and the salmonid population increasing in both frequency and density as the pH increases from less than 5.0 to more than 6.5. Only 28 per cent of the streams with pH < 5.5 contained salmon. Although eels are considerably more tolerant of acid conditions than are salmonids, they were twice as common in streams of pH > 6.0 compared with streams of pH < 5.5.

Chemical monitoring of some 500 lakes in Norway by Wright and Snekvik (1978), together with estimates of their fishery status in the 1970s, revealed that the latter was largely independent of pH for 75 per cent of the lakes with pH < 5.1, about 50 per cent of the lakes being fishless. Only when the pH exceeded 5.1 did the fishery status improve,

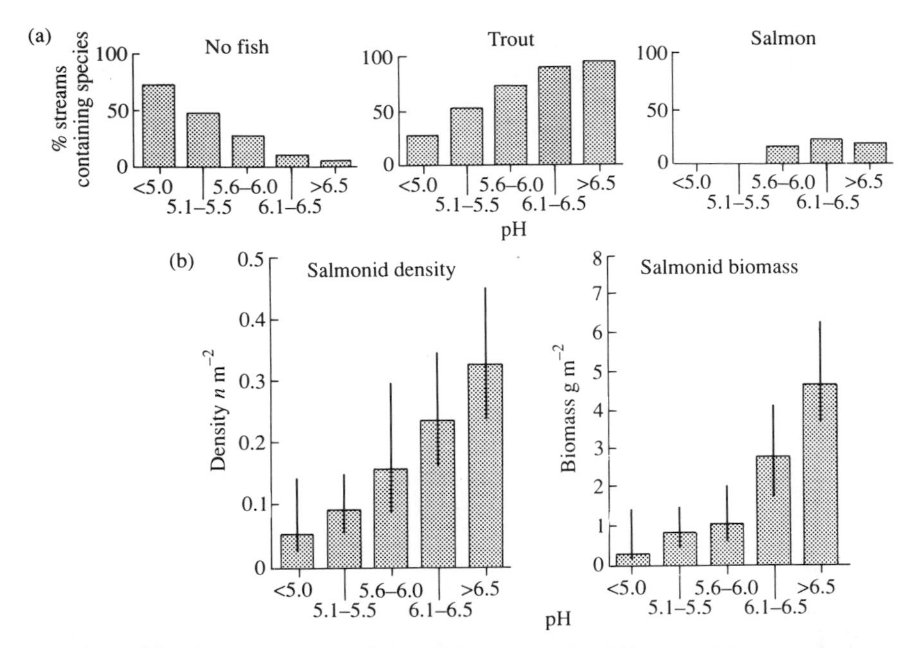

Fig. 8.1 (a) The percentage of Scottish streams in different pH categories that contain salmon and trout. The percentage of fishless lakes is also shown. (b) Salmonid population density and biomass in different pH categories. (From Turnpenny 1989, p. 56, by permission of the Cambridge University Press.)

the two most important determinants being [Ca^{2+}] and pH, in that order. Figure 8.2 shows that the majority of fishless lakes were those with [Ca^{2+}]/[H^+] $<$ 3, whereas most of the lakes with good fisheries had [Ca^{2+}]/[H^+] $>$ 4. Because of cross-correlations between the variables, these early surveys did not reveal a clear-cut effect of aluminium on the fisheries.

A follow-up survey, involving about 1000 lakes, by Wright and Henriksen in 1986, see Henriksen *et al.* (1990) revealed that the proportion of fishless lakes had increased markedly during the intervening 12 years. In southernmost Norway, about half of the 177 lakes which contained fish in the early 1970s were barren in 1986, reproduction probably having ceased a decade earlier. This was accompanied by a decline in the concentrations of base cations, Ca^{2+} and Mg^{2+} and an increase in aluminium. In eastern Norway the trends were not so clearly marked because several species (e.g. perch) besides brown trout were involved and because the lakes were often rich in humus, which forms organic compounds of aluminium which are not toxic to fish. In western Norway there was evidence that the water quality had improved during the past decade, owing partly to reduced acidic deposition and partly to fewer enhanced acidic episodes.

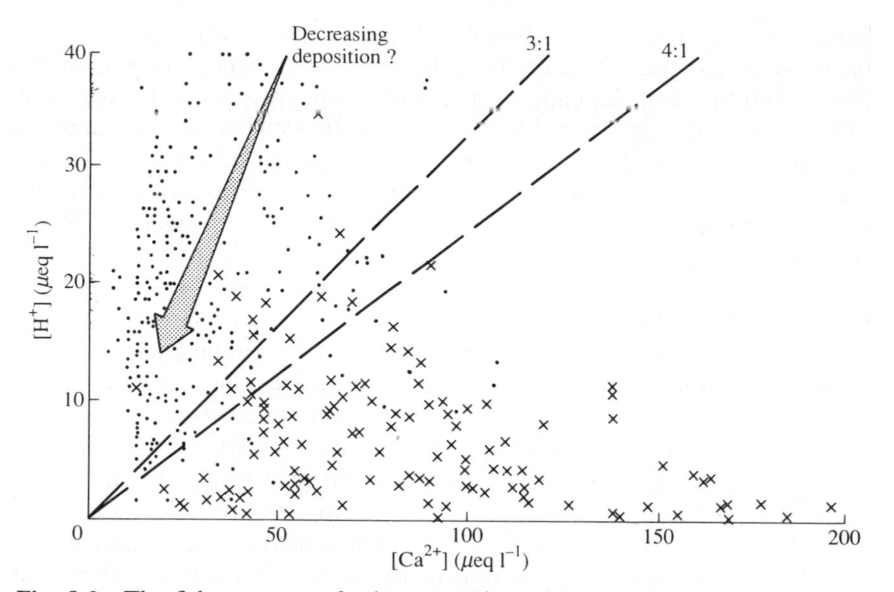

Fig. 8.2 The fishery status of a large number of Norwegian lakes in relation to their acididty (H^+ concentration) and the calcium concentration: ·, no fish; ×, good fish. (From Chester 1984, by permission of The Royal Society.)

A worrying aspect concerning the future of fish populations in southern Norway is the recent tendency for decreasing concentrations of sulphate in the lakes to be offset by increasing concentrations of nitrate (see Section 7.6), with little or no reduction in acidity.

A more rigorous statistical analysis of these Norwegian data on fishery status and water chemistry, to include a larger number of variables, has recently been published by Muniz and Walløe (1990). Bivariate analysis confirmed the previously observed correlations between brown trout survival and each of the independent variables. However, in different step-wise multi-variate analyses involving 14 chemical and physiographic variables, fish survival appeared to depend only on pH, concentration of inorganic aluminium [Al_i], and lake altitude, in that order. The Ca^{2+} ion concentration did not emerge as a significant factor and led the authors to infer that the low concentrations of calcium normally present in Norwegian lakes were neither harmful to fish nor protected them from the deleterious effects of aluminium.

These conclusions have been strongly challenged by many scientists working on this problem, both in the field and with controlled laboratory experiments (see below). The apparent contradiction between the Muniz–Walløe analysis and the earlier results of Wright and Snekvik may perhaps be attributed, at least in part, to the fact that the former was based on the presence or absence of brown trout, whereas the latter included mixed fish populations including perch, which require higher concentrations of calcium and magnesium to survive than do brown trout. The apparent disagreement between the Muniz–Walløe analysis and the laboratory experiments, which point strongly to the protective effects of calcium, may be partly due to the fact that the experiments were mostly restricted to short periods and to low values of $pH < 4.5$, whereas 50 per cent of the lakes in southern Norway have $pH > 5.0$.

Even so, it is difficult to believe that lake altitude *per se* can be a significant determinant of fishery status but it may well be a partial surrogate for one or more other parameters. The fact that high altitude lakes have sparse trout populations may be partly due to the incidence of low temperatures, heavy snowfall, and low calcium concentrations. The speciation and toxicity of aluminium is temperature sensitive, while melting snow produces highly acidic episodes and variable water chemistry that are poorly buffered by calcium and other basic cations.

There is evidence to show that fish, including trout, can survive with very low concentrations of calcium in the absence of acid or aluminium stress. Conversely, the addition of inorganic aluminium in concentrations greater than 50 $\mu g\ l^{-1}$ does appear to reduce the growth rates of yearling trout, but this can be counteracted by adding calcium.

The questions of whether calcium can protect against high acidity and aluminium might be resolved by monitoring some carefully selected lakes low in calcium and with declining trout populations, enhance the calcium concentrations without changing the pH or $[Al_i]$, and then determine their rate of recovery relative to nearby control lakes. Since the calcium concentration may be particularly critical at the spawning stage, it might be better to add the calcium salt to the spawning stream rather than to the whole lake.

It is important to resolve this question of the extent to which calcium may ameliorate the toxic effects of acidity and aluminium, since it has major implications concerning the rate of recovery of fish populations following reductions in acid deposition and the efficacy of liming strategies.

Although an analysis of the presence or absence of fish may indicate the basic pattern of extinction in relation to soil and water chemistry, the effects of such chemical changes are reflected in changes in mortality, recruitment, and growth. Therefore an analysis of the structure and dynamics of a population which incorporates the current health of the stock is more informative and helpful in predicting its future evolution in changing environmental conditions.

Such an analysis was carried out in the SWAP by Bravington *et al.* (1990). Data from two Norwegian sources, one a long-term study of a continuous liming programme, and the other a study of some 100 representative lakes across the country, yield characteristic age compositions for brown trout populations undergoing responses to differing physical and chemical conditions. There are two main responses: senescence, which occurs when partial or complete recruitment failures lead to a population with a preponderance of large, old fish which may not appear to be at risk, and juvenilization, when increased post-spawning mortality leads to a population having a large proportion of immature fish with large year-class variability. These responses have been related to physical and chemical characteristics of the environment, notably lake area, pH and aluminium and organic content of the water.

The observed age compositions are fitted to simple models incorporating changing mortality and recruitment, which are then projected forwards to predict longer-term effects on population structure in response to environmental change. The median times to extinction for senescent populations of trout are generally shorter (less than 10 years) than for juvenilized population (20–40 years), even in lakes of similar water quality, but the underlying reasons are not clear.

8.2 The effect of acidic episodes on fish mortalities

There are several reports in the literature of major fish kills apparently following episodes of unfavourable water chemistry (e.g. high acidity accompanied by high concentrations of aluminium) but rather few of these are supported by actual measurements of chemical species, flow rates, etc., which would require frequent, if not continuous, sampling and this is rarely available. As described in Chapter 4, acidic episodes may have several causes, are associated with heavy rains after a dry spell and with snow melt, both of which give rise to high flow rates in the streams. Although they may last anything from a few hours to a few weeks, changes in pH and other parameters can be very rapid and peak values may be sustained for only a few hours or even minutes. Thawing of snow during the day and refreezing at night may give rise to a series of short episodes. Figure 8.3 shows the changes in the chemistry of a high altitude stream in central Sweden during a spring snow melt which resulted in 100 per cent mortality of brown trout following a fall in pH from 6.0 to 4.6, sharp increases in aluminium, manganese and iron, and a sharp fall in calcium due to the dilution by melt water passing over the frozen soil and entering the stream directly. Henriksen *et al.* (1988) reports fish kills associated with acidic episodes in the River Viksdalselva every year from 1981 to 1984. In 1985 there was little snow, no major acid episodes, and no fish kills. In 1986 fish kills were recorded in both the river and in experimental tanks flushed with river water. The majority of pre-smolt salmon died after 12 days with pH 5.1–5.3, $[Al_i] = 50–60\ \mu g\ l^{-1}$, $[Ca] = 0.55–$

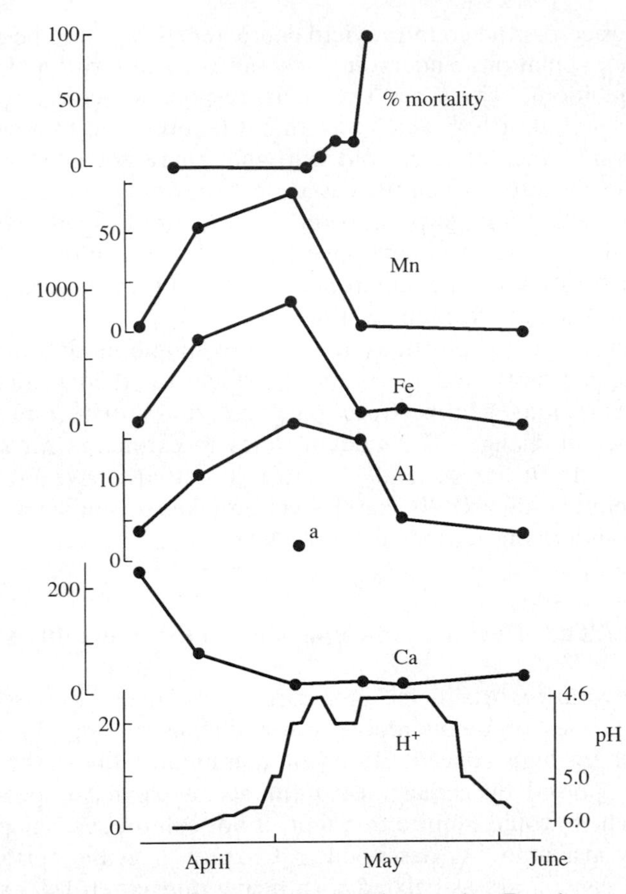

Fig. 8.3 Changes in the water quality in Bjursvasslan Brook, central Sweden, during snow melt in spring 1982, with the cumulative mortality of experimentally exposed brown trout. The thaw started in mid-April and maximum run-off occurred in mid-May. The concentrations of the various chemical species are in micromoles per litre. (From Reader and Dempsey 1989, p. 75, by permission of the Cambridge University Press.)

0.75 mg l^{-1}. Apparently, aluminium was easily mobilized from the river bed and $[Al_i]$ increased rapidly with falling pH. This was tested by artificially acidifying a small stream during low flow. Up to 70 per cent of the added acid was exchanged with aluminium on the stream bed, releasing a pool of exchangeable aluminium ions which proved toxic to the salmon at low pH.

Harriman *et al.* (1990) studied the relationship between stream flow rates, water chemistry (especially pH, [Ca] and [Al]), and the survival of salmon eggs in the Allt a'Mharcaidh stream in Scotland. They also studied the changes in the numbers and diversity of invertebrates. They found that pH, alkalinity, [Ca], [Na], and [Si] to be negatively correlated with stream flow, TOC and [Al] to be positively correlated, and [SO_4], [NO_3], and [Cl] to be uncorrelated. High [Na], [Si], and [Ca] at low flows suggest weathering of Na–Si minerals (e.g. albite) releasing Na^+ ions, and of Ca–Si minerals (e.g. anorchite) to release Ca^{2+} ions. High flows produce high [Al] and [H^+] because these are associated with heavy rain and snow melt, releasing aluminium, H^+ and organic acids from upper layers of the soil through which most of the percolate flows in these conditions. The fact that high flows produce a high output of SO_4 without the concentration being increased, suggest a large storage of SO_4 in the upper horizons.

During acidic episodes, following heavy rains or snow melt, when the pH dropped below 5.5 and [Ca^{2+}] fell below 50 μeq l^{-1} (1 mg l^{-1}), there were significant mortalities of salmon fry. The survival of both eggs and alevins was enhanced by about one third when these were protected by packs of limestone which raised [Ca^{2+}] to around 60 μeq l^{-1}, compared with values of around 40 μeq l^{-1} in the gravel, the pH increasing by only approximately 0.1 unit. Since concentrations of labile aluminium were low and did not exceed 13 μg l^{-1} even during episodes, aluminium was unlikely to have been a significant influence upon egg and fry survival.

An important factor in salmonid survival is the frequency and timing of acidic episodes relative to the stage in the life cycle. The first episode may induce sublethal physiological stress from which the fish may take several weeks to recover, but further episodes soon after the first may cause mortalities.

Freshly fertilized eggs are sensitive to changes in water chemistry but become less sensitive once the egg membrane has hardened. However, as soon as the eggs hatch and the alevins are exposed directly to the stream water, they become much more susceptible to abrupt reductions in pH.

According to Potts *et al.* (1990), adult salmon returning from the sea to their home fresh-water rivers, gradually adjust to the new conditions by activating their 'sodium pumps', which involves replacement of an epithelium on the gills so that they can take up sodium from the dilute medium. Even so, sodium loss remains high for several hours and results in a fall in the sodium concentration in the blood. Potts *et al.* find that if salmon are rapidly transferred from sea water, which has a pH of approximately 8, into acid water of pH 5, sodium uptake is inhibited by 90 per cent, resulting in a marked fall in the blood concentration. Exposure for 6 hours to water of pH 5 containing 20 μmol l^{-1} of aluminium severely inhibits sodium intake for at least 24 hours after return to neutral water. The authors conclude that exposure to two such short acid, high aluminium episodes 24 hours apart would reduce sodium in the blood to lethal

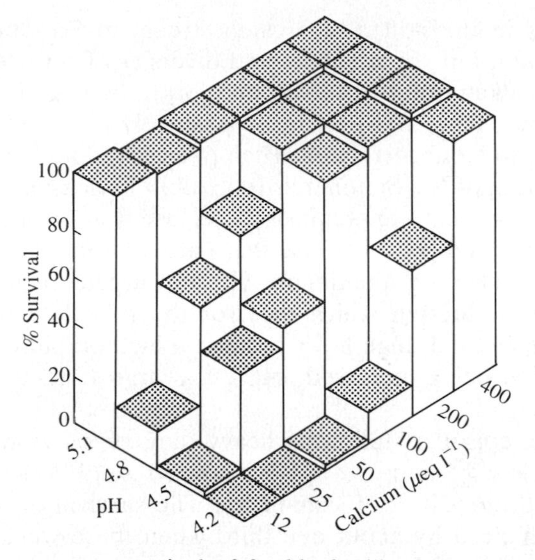

Fig. 8.4 The percentage survival of freshly fertilized brown trout eggs after 8 days' exposure to range of pH and calcium concentrations. (From Brown and Sadler 1989, p. 37, by permission of the Cambridge University Press.)

levels, and that such occurrences might well account for major fish kills observed in rivers such as the Esk and Duddon in Cumbria.

8.3 Studies of fish mortalities in laboratory experiments

Since data from natural streams are rather sparse and difficult to obtain, much of the existing information concerning the relationship between fish survival and water chemistry is based on experiments in laboratory tanks containing synthetic chemical solutions. Unfortunately, in many early experiments the chemical composition was not adequately controlled and was often far removed from that of natural waters susceptible to accidification, and this led to conflicting results. In such experiments it is important to control the pH carefully because the solubility of aluminium, its speciation, and hence its toxicity are very sensitive of pH.

The importance of calcium in determining the survival of freshly fertilized brown trout eggs at low pH was clearly demonstrated by Brown (1982), as illustrated in Fig. 8.4 Survival after 8 days was 100 per cent at pH 5.1 irrespective of calcium concentration down to 12 $\mu eq\ l^{-1}$, and irrespective of pH down to 4.2 if $[Ca^{2+}]$ = 400 $\mu eq\ l^{-1}$, whilst at low pH (4.2) and low $[Ca^{2+}]$, i.e. less than 50 $\mu eq\ l^{-1}$, survival was nil. The interrelated effects of pH, calcium, and inorganic aluminium on the survival of brown trout yolk-sac fry are shown in Fig. 8.5.

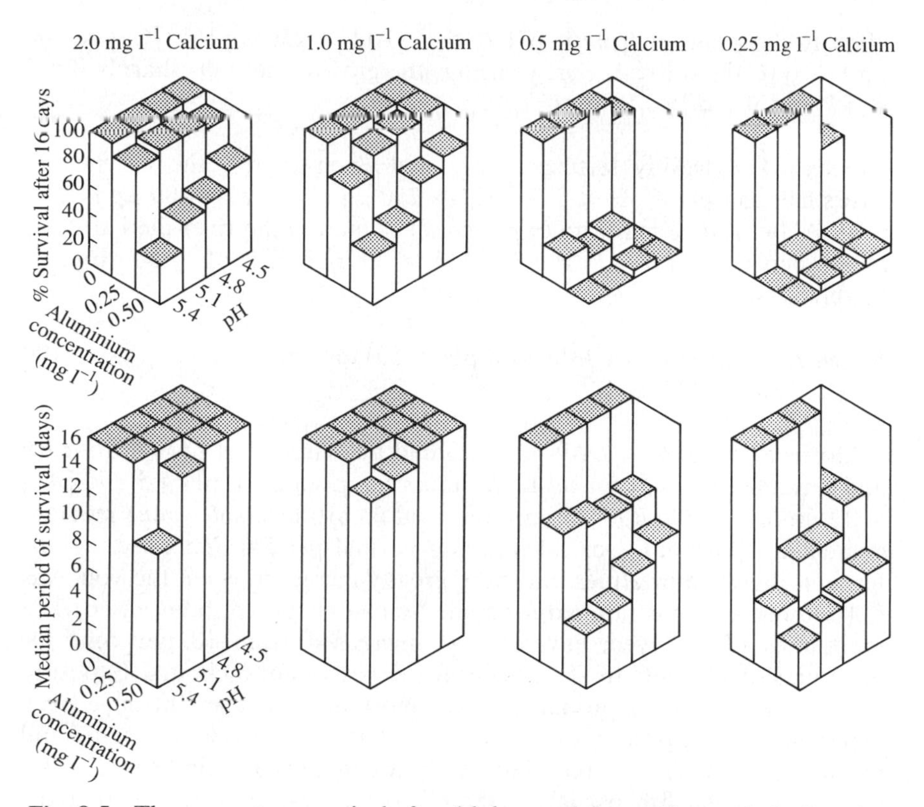

Fig. 8.5 The percentage survival after 16 days and the median period of survival of brown trout fry after exposure to a range of pH, calcium, and aluminium concentrations. (From Brown 1983, by permission of Dr D.J.A. Brown.)

In the absence of aluminium, no deaths occurred after 16 days with pH between 4.8 and 5.4. Deaths occurred at pH 4.5 only for the lowest calcium concentrations of 0.25 mg l^{-1} and 0.5 mg l^{-1}. With aluminium added as chloride in concentration 0.25 mg l^{-1}, there was high survival if [Ca] > 1 mg l^{-1}, but a sharp drop ensued if this fell below 0.5 mg l^{-1} or the [Ca]/[Al_i] ratio fell below 2. When [Al_i] was increased to 0.5 mg l^{-1}, the mortality was high even with [Ca] = 2 mg l^{-1} and [Ca]/[Al] = 4, and was higher at higher pH values because of the increased solubility of aluminium.

Evidence concerning the effects of pH, calcium, and aluminium on the survival of salmonids at various stages of their life history may be summarized as follows.

Brown trout. For freshly fertilized eggs the survival after 16 days falls sharply if pH < 5.0 and [Ca] < 1 mg l^{-1}. For swim-up fry the survival after 16 days falls sharply if pH < 5.4, $[Al_i] > 250\ \mu g\ l^{-1}$, $[Ca] < 500\ \mu g\ l^{-1}$ so $[Ca]/[Al_i] < 2$. For, yearlings the growth rate falls sharply if pH < 5.2 and if $[Al_i] > 50\ \mu g\ l^{-1}$.

Salmon. For freshly fertilized eggs and swim-up fry, the survival in rivers falls sharply if pH < 5.5, $[Al_i] > 200\ \mu g\ l^{-1}$, $[Ca] < 400\ \mu g\ l^{-1}$, so $[Ca]/[Al_i] < 2$. The majority of pre-smolt salmon in the river died after 12 days when the pH was 5.1–5.3, $[Al_i] > 50\ \mu g\ l^{-1}$, $[Ca] < 500\ \mu g\ l^{-1}$, $[Ca]/[Al_i] < 10$.

In general. Streams are fishless if pH < 5.0 for long periods, $[Al_i] > 100\ \mu g\ l^{-1}$, $[Ca] < 200\ \mu g\ l^{-1}$.

Morris and Reader (1990) have studied the survival of brown trout fry and juveniles exposed for up to 78 hours to episodes of pH 4.5 and $[Al_i] = 12\ \mu mol\ l^{-1}$ ($320\ \mu g\ l^{-1}$) impressed on an artificial soft-water medium containing $20\ \mu mol\ l^{-1}$ of calcium and normal pH 5.6. Yolk-sac fry suffered negligible mortalities but these greatly increased when the yolk was fully absorbed and continued for some time after the episode ceased. The mortalities of 1–2 year juvenile fish increased from 10 per cent for exposures of 12 hours to 75 per cent for exposures of 60 hours. Exposure to successive episodes produced lower mortalities than might have been expected from exposure to a single episode of equivalent length, and mortalities declined as the interval between episodes increased. This suggests that the fish partially adapted to the first episode and that the period of recovery between episodes is an important factor in determining survival. Juveniles surviving a 30-hour episode showed a significant lowering of sodium and chloride in the blood plasma but these recovered to near normal values after 5 days, although the gills showed signs of damage.

Morris and Reader have also investigated the effects of episodic changes in pH and aluminium on the sodium balance and respiration in brown trout using a computer-controlled rig that provides a continuous flow of water, the flow rate, chemistry, and temperature of which are carefully monitored and controlled. These parameters can be readily changed so that small juvenile trout can be subjected to acid and/or aluminium episodes of predetermined amplitude and duration, and then allowed to recover. The sodium balance of the fish is determined from measurements of the influx and outflux of sodium using radioactive ^{24}Na isotopes as a tracer, and the plasma sodium content is measured at the end of the experiment to give the total sodium loss. The rate of oxygen consumption is also measured.

In one experiment, when the pH was allowed to fall to 4.5 and $[Al_i]$ to increase to $150\ \mu g\ l^{-1}$ over 6 hours, remain at those levels for 6 hours, and then return over the following 6 hours, the fish showed a rapid loss of sodium during the build-up phase, mainly as the result of increased outflux, which gradually declined as the episode progressed. The sodium loss was much attenuated with high levels of calcium in the water.

The main effect of aluminium was to reduce the influx of sodium, which was almost completely inhibited by the end of the plateau phase, but recovered 50 per cent within 24 hours after the end of the episode. Recovery from acid alone is faster than from a combined acid–aluminium episode, when aluminium concentrations as low as 50 $\mu g\ l^{-1}$ can produce marked effects.

Figure 8.6 shows the effects of a combined 24-hour episode of acid and aluminium intended to give 'plateau' values of pH 4.5 and $[Al_i] = 10\ \mu mol\ l^{-1}$. Influx of sodium is reduced and outflow is much increased during the rising phase of the episode but both gradually recover during the falling phase and for some time after the end of the episode. The overall loss of body sodium is small during acid episodes but increases with increasing levels of aluminium attained during combined episodes. In the case represented by Fig. 8.6, it was about 20 per cent.

The fish showed no significant changes in oxygen consumption during most of the short episodes in these experiments.

This work could be usefully extended to study mortalities arising from episodes of longer duration and intensity and as may occur in natural streams.

8.4 Effects on other aquatic biota

Once lakes and streams become acidified, considerable changes may occur, not only in fish populations but in the composition of the whole ecosystem. There is a good deal of evidence that acidification is accompanied by decreased diversity in phytoplankton, zooplankton, and benthic invertebrates. Reductions of invertebrate populations are likely to have serious consequences for fish because they are part of the latter's food supply. For a recent detailed review see Herrmann (1990).

Most crustaceans, molluscs, and mayfly nymphs are vulnerable to acid conditions, whereas other species, such as dragonfly and chironomid larvae, and some species of stoneflies, are less sensitive.

Ion regulation in mayflies, daphnids, and some stoneflies is adversely affected by low pH and increasing levels of aluminium and heavy metals, both of which cause net losses of sodium and other ions which are especially serious in the moulting stage. Calcium metabolism is often upset at low pH, especially in crayfish and molluscs, so that the addition of calcium often aids survival.

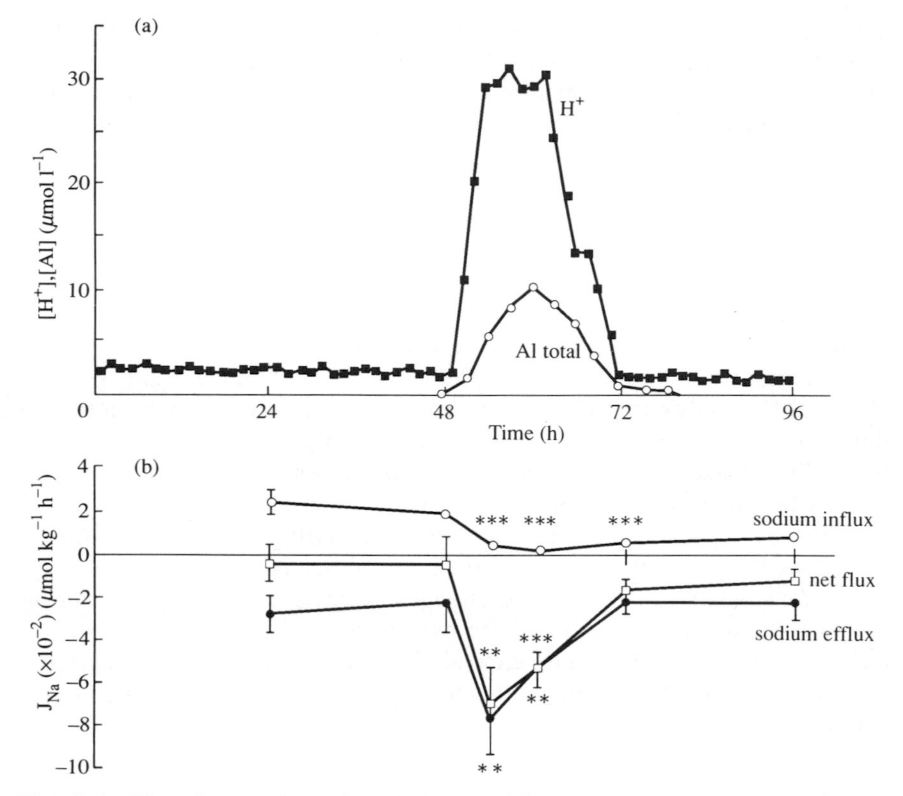

Fig. 8.6 The changes in sodium balance of brown trout when subjected to a short combined acid (pH 4.5) and aluminium episode. (From Morris and Reader 1989, p. 362, by permission of the Cambridge University Press.)

Besides impairing the physiological functions of certain species, acidic conditions may also damage or interrupt the food chain and thereby seriously impair the survival of animals at higher trophic levels, such as fish. Thus insects such as mayfly nymphs, which graze on algae on the stream bed, and crustaceans (shredders), which feed on decaying plant detritus, are adversely affected because the acids kill the bacteria responsible for decomposing these foodstuffs.

The composition of a lake ecosystem may be significantly influenced by the presence of mosses and fungal mats which modify the exchange of nutrients such as phosphorus between the sediments and the water body. Furthermore, mosses may serve as reservoirs for precipitated aluminium that can be mobilized if the pH falls during, for example, a storm or snow-melt episode.

8.5 The physiology of ion regulation and salt balance in fishes

Blood plasmas of freshwater fishes contain around 150 meq l^{-1} Na^+ and 130 meq l^{-1} Cl^- together with much lower concentrations of other ions such as Ca^{2+}, K^+, and HCO_3^-. In contrast, soft fresh waters usually contain less than 0.1 meq l^{-1} NaCl and much lower concentrations of other ions. Most species of fish, including all salmonids, take up salt from the dilute medium to balance the loss by osmosis and in urine. Most of the loss takes place across the body wall, and especially across the gills, where the blood plasma and the external medium are separated by only a thin layer of respiratory epithelium where the blood is oxygenated. The gills are also mainly responsible for salt intake.

Sodium uptake is driven by the enzyme Na-K-ATPase located in the gill membrane which, in effect, pumps Na^+ ions from the dilute external medium into the blood across the epithelial membrane. The mechanism of this ion pump is not well understood, but some of its characteristics, inferred largely from measurements of ion exchange and transport, are described below. Sodium uptake requires either simultaneous transport of an anion, or the export of another cation in order to maintain charge balance. The exchange cation could be NH_4^+ but is more likely to be H^+. The transport of Cl^- ions is more obscure but they are probably exchanged for HCO_3^- and OH^- ions. It appears that the gills are able to transport, independently and through ion-specific channels in the outwardly facing epithelial membrane, the three electrolytes, Na^+,Cl^-, and Ca^{2+}, and possibly also K^+, each ion having its specific controlling ATPase. Separate from this, there is a passive and less specific outward leakage of electrolytes through the intercellular junctions, the loss rate being proportional to the rate of uptake.

The efflux, and hence the loss, of sodium from the blood occurs mainly by simple diffusion across and between the cells, and is dependent on the permeability of the epithelium. This is controlled by the potential developed across the membrane by the preferential adsorption of ions on both sides and the diffusion gradients of all ions present in its neighbourhood. In neutral waters the dominant ions are Na^+ and Cl^- but, in acid waters, H^+ ions play an important role because the gills are highly permeable to them. In the latter case, any imbalance between the efflux of Na^+ and Cl^- ions is compensated by an influx of H^+ ions. At low pH, the efflux of Na^+ is increased; in brown trout, it may more than double as the pH falls from 7.0 to 4.0 because the potential difference across the gill surface is increased owing to its high permeability to H^+ ions and because H^+ ions displace Ca^{2+} on the membrane and increase its permeability to Na^+ ions.

Turning now to the influx of Na^+ ions from the dilute medium, the ion pump works most efficiently when the Na^+ concentration in the water is much greater than the H^+ concentration. Since the system essentially exchanges Na^+ for H^+ ions, the transfer of Na^+ is bound to be reduced at low pH, when the H^+ ion concentration in the water is increased. Several mechanisms may be involved, for example competition between Na^+ and H^+ ions for for the transport sites on the gill membrane, damage to the enzyme ATPase at low pH, and leakage in the ion pump resulting in reduced Na^+ influx. In summary, low pH enhances the efflux of sodium, and reduces the influx, so that the uptake by the fish is reduced on both counts, and increasingly so as the pH falls below 5.0. At pH < 4.0, influx virtually ceases and efflux is greatly stimulated.

The effects of low pH on the fluxes of Cl^- ions involve rather difficult measurements and have been little studied. However, in general, the results are similar to those for Na^+. If the Na^+ concentration in the blood falls, that of Cl^- soon follows. If $[Na^+]$ falls and H^+ ions enter the blood, HCO_3^- is converted to CO_2, $[HCO_3^-]$ falls and Cl^- uptake is correspondingly reduced.

The permeability of the gill membrane to both Na^+ and H^+ ions, and hence the sodium balance of the fish, is strongly influenced by quite low concentrations of Ca^{2+} and inorganic aluminium ions in the water. Since the outer surface of the gill membrane is negatively charged in acid water, it has an affinity for metallic cations. Binding of Ca^{2+} ions on the surfaces of the transport channels in the membrane reduces its permeability to cations, especially to H^+ and Na^+ ions, lowers the membrane potential, and reduces both the Na^+ loss and the influx of H^+. Experiments on rainbow trout have shown that as the $[Ca^{2+}]$ in the water increases, $[Na^+]$ and $[Cl^-]$ in the blood plasma increase and the acidity of the blood falls.

Certain species of aluminium ions, notably Al^{3+} and the inorganic hydroxides, $AlOH^{2+}$ and $Al(OH)_2^+$ act synergistically with H^+ ions to decrease sodium uptake and increase sodium loss, whilst Ca^{2+} ions help to protect fishes from both effects. The different species of aluminium ions affect sodium efflux and influx to different degree depending upon the pH. Stimulation of sodium efflux by aluminium is greatest in the pH range 5.5–5.2 and is thought to be largely due to $Al(OH_2^+$ ions, whilst inhibition of sodium influx is greatest at pH 4.5–4.0, owing mainly to Al^{3+} and $Al(OH)^{2+}$. Aluminium in a concentration of 200 $\mu g\ l^{-1}$ (7.5 μM) and at pH 5.0 reduced the activity of the gill Na-K-ATPase by 25 per cent in salmon and rainbow trout during experiments lasting over 2–5 days, and resulted in a marked reduction in $[Na^+]$ and $[Cl^-]$ in the blood plasma. The interaction between aluminium and calcium ions is far from clear. There is evidence that while aluminium ions displace Ca^{2+} ions from the microsomes of gill cells, they have no marked effect on Ca^{2+} ions bound on the gill surface.

More detailed discussions of these subjects may be found in Wood (1989) and in Potts and McWilliams (1989).

8.6 References

Bravington, M.V., *et al*. (1990). In the *surface waters acidification programme*, (ed. B.J. Mason) pp. 467–76. Cambridge University Press.

Brown, D.J.A. (1982). *Water, Air and Soil Pollution*, **18**, 343–51.

Brown, D.J.A. (1983). *Bulletin of Environmental Contamination and Toxicology*, **30**, 582–7.

Brown, D.J.A. and Sadler, K. (1989). In *Acid toxicity and aquatic animals*, (ed. R. Morris, *et al*.), pp. 31–44. Cambridge University Press.

Chester, P.F. (1984). *Proceedings of the Royal Society of London, Series B,* **305**, 564–5.

Harriman, R., *et al*. (1990). In *The surface waters acidification Programme*. (ed. B.J. Mason), pp. 343–545. Cambrdige University Press.

Henriksen, A., *et al*. (1988). In *Proceedings of the mid-term conference on SWAP*, (ed. B.J. Mason), pp. 163–76. The Royal Society.

Henriksen, A., *et al*. (1990). In *The surface waters acidification programme*, (ed. B.J. Mason), pp. 199–212. Cambridge University Press.

Herrmann, J. (1990). In *The surface waters acidification programme*. (ed. B.J. Mason), pp. 383–96. Cambridge University Press.

Morris, R. and Reader, J.P. (1990). In *The surface waters acidification programme*, (ed. B.J. Mason), pp. 357–68. Cambridge University Press.

Muniz, I.P. and Walløe, L. (1990). In *The surface waters acidification programme*, (ed. B.J. Mason), pp. 327–42. Cambridge University Press.

Potts, W.T.W., *et al*. (1990). In *The surface waters acidification programme*, (ed. B.J. Mason), pp. 369–82. Cambridge University Press.

Potts, W.T.W. and McWilliams, P.G. (1989). In *Acid toxicity and aquatic animals*, (ed. R. Morris, *et al*.), pp. 201–20. Cambridge University Press.

Reader, J.P. and Dempsey, C.H. (1989). In *Acid toxicity and aquatic animals*, (ed. R. Morris *et al*.), pp. 67–83. Cambridge University Press.

Turnpenny, A.W.H. (1989). In *Acid toxicity and aquatic animals* (éd. R. Morris, *et al*.), pp. 45–65. Cambridge University Press.

Wood, C.M. (1989). In *Acid toxicity and aquatic animals*, (éd. R. Morris, *et al*.), pp. 125–52. Cambridge University Press.

Wright, R.F. and Henriksen, A. (1986).

Wright, R.F. and Snekvik, E. (1978). *Verhandlungen Internationale Verein Theoretische Angewandte Limnolnologie*, **20**, 765–75.

FURTHER READING

Mason B.J. (ed.) (1990). The *surface waters acidification programme*, pp. 327–430. Cambridge University Press.

Brown, D.J.A. and Lynam, S. (1981). *Journal of Fish Biology*, **19**, 205–11.

Morris, R., Taylor, E.W., Brown, D.J.A., and Brown, J.A. (eds.) (1989). *Acid toxicity and aquatic animals*. Cambridge University Press.

9
Models of acidification of surface waters

9.1 Introduction

A number of numerical models to simulate, and perhaps predict, changes in water flow and chemistry, in both the soil and the stream, in response to changes in inputs of water and chemical deposition, have been develped in recent years. Prominent among these are the MAGIC model (Model of Acidification of Groundwaters in Catchments) developed by Cosby *et al.* (1985) in the USA and the BIRKENES model developed in Norway which have been applied to a variety of catchments in North America, Scandinavia, and the UK.

A satisfactory model should characterize the principal hydrochemical processes, account for the changing levels of inputs, and provide good estimates of past, present and future changes in soil and water chemistry. It should also be transferable in the sense that it can be readily applied to a wide range of catchments in differing pollution climates with differing land-use regimes and differing soils and parent geologies. No model, at present, meets all these requirements. All are necessarily gross simplifications of what happens in a real catchment and are calibrated with, or tuned to fit, observed data that are often lacking in spatial and temporal coverage, in quality and continuity.

9.2 The MAGIC model

9.2.1 MODEL STRUCTURE

The MAGIC model is designed to simulate *long-term* changes in surface-water chemistry and to infer what changes may occur in future in response to various assumptions about future levels of acid deposition. It is basically a chemical model driven mainly by changes in sulphate inputs. The incoming precipitation is assumed to react chemically with the soil and emerge as stream water with no losses by run-off or storage as ground water.

The original version of the model is governed by 24 equations which represent equilibrium between chemical reactions in the soil and soil/stream water. The processes considered are listed below:

(1) the generation of alkalinity by dissociation of carbonic acid under the high CO_2 pressures that exist in the soil and subsequent exchange of H^+ ions for base cations;

(2) anion retention by soils, e.g. by adsorption/desorption of SO_4^{2-} ions;

(3) adsorption and exchange of base cations and aluminium cations (Al^{3+}, $Al(OH)_2^+$, $AlOH^{2+}$, . . .) for H^+ ions by soils;

(4) weathering of minerals to produce base cations, the weathering rates assumed to be constant;

(5) control of Al^{3+} concentrations by assumed equilibrium with a single solid phase, e.g. $Al(OH)_3$ — gibbsite.

The chemical conditions are assumed to be constant throughout the soil profile, and the key reactions are taken to be

(a) cation exchange between the percolate and soil involving Al^{3+}, Ca^{2+}, Mg^{2+}, K^+, Na^+ (four equations), changes in cation exchange capacity (1) and in base saturation (1), giving a total of *six equations*,

(b) inorganic aluminium reactions involving hydroxy, fluorine, and sulphate species, *13 equations*,

(c) dissolved inorganic carbon reactions involving dissolved CO_2 to produce HCO_3^-, OH^-, CO_3^{2-}, H^+, *four equations*,

(d) ionic charge balance, *one equation*,

making a total of *24 equations* that contain 33 variables and 21 parameters.

Variables	Parameters
Concentrations of 12 ionic species and of 13 aluminium species	Aluminium solubility constant, one parameter
Five exchangeable cation fractions, E_{AL} . . .	Activity constants for 12 aluminium species, 12 parameters
Base saturation of the soil	Solubility constants for CO_2, three parameters
pCO_2 in the soil	Dissociation constant for water, one parameter
Concentration of dissolved CO_2	Selectivity coefficients for cation exchange of Na^+, K^+, Ca^{2+}, Mg^{2+}, four parameters
Total 33	Total 21

The chemical reactions will now be formulated in more detail.

9.2.2. INORGANIC CARBON REACTIONS IN SOIL WATER

The soil water is assumed to be in equilibrium with CO_2, existing at elevated pressure in the soil matrix, to form dissociated carbonic acid, i.e.

$$CO_2 + H_2O = HCO_3^- + H^+ \quad (9.1)$$

or

$$= CO_3^{2-} + 2H^+$$

represented by

$$\{CO_2(aq.)\}/p(CO_2) = K_{CO_2}, \quad \{HCO_3^-\}\{H^+\}/\{CO_2(aq.)\} = K_{CO2,3}$$

$$\{CO_3^{2-}\}\{H^+\}/\{HCO_3^-\} = K_{CO3,3}, \quad \{H^+\}/\{OH^-\} = K_w$$

where { } denotes activities, and involving four dissociation constants.

9.2.3 SOLID AND AQUEOUS PHASE ALUMINIUM REACTIONS

It is assumed that the concentrations of Al^{3+} and other hydroxy species such as $Al(OH)^{2+}$, $Al(OH)_2^+$, ... in the soil water are in equilibrium with the solid phase $Al(OH)_3$ (gibbsite) and that the reactions such as

$$3H^+ + Al(OH)_3(s) \rightleftharpoons Al^{3+} + 3H_2O \quad (9.2)$$

occur instantaneously as the concentrations of H^+ and Al^{3+} vary. The equilibrium concentrations (activities) of Al^{3+} and H^+ ions in solution are then represented by

$$\{Al^{3+}\}/\{H^+\}^3 = K_{Al}, \quad (9.3)$$

K_{Al} being the aluminium solubility constant.

The model includes additional reactions involving hydroxy, fluorine, and sulphate species, two of which are represented by

$$\{Al(OH)^{2+}\}\{H^+\}/\{Al^{3+}\} = K_{Al_2} \quad (9.4)$$

and

$$Al^{3+} + 2H_2SO_4 \rightarrow Al(SO_4)_2^- + 4H^+ \quad (9.5)$$

yielding

$$\{Al(SO_4)_2^-\}/\{Al^{3+}\}\{SO_4^{2-}\}^2 = K_{Al_{12}} \quad (9.6)$$

The aluminium solubility constant is not a true thermodynamic constant but varies from soil to soil and has to be estimated from field data or by appropriate calibration of the model.

The 12 Al_i activity constants are assumed not to vary from catchment to catchment (although they are temperature dependent) and to have been determined by experiment.

9.2.4 CATION EXCHANGE BETWEEN SOIL AND SOIL WATER

Cation exchange between the soil and the percolate involving Al^{3+}, Ca^{2+}, Mg^{2+}, K^+, and Na^+ is represented by equations of the form:

$$2Al^{3+} + 3CaX_2 = 3Ca^{2+} + 2AlX_3 \quad (9.7)$$

$$Ca^{2+} + 2\,NaX = 2Na^+ + CaX_2 \quad (9.8)$$

$$Mg^{2+} + 2\,NaX = 2Na^+ + MgX_2 \quad (9.9)$$

$$K^+ + NaX = Na^+ + KX \quad (9.10)$$

where CaX_2, AlX_3, MgX_2,... denote the cations adsorbed on the soil, and Al^{3+}, Ca^{2+}, Mg^{2+}, ... denote those in the soil water.

The equilibrium concentrations (activities) of the ions for these reactions are expressed in terms of activity coefficients of ion species in the water and equivalent fractions E of ions adsorbed on the soil, e.g.

$$\frac{\{Ca^{2+}\}^3\, E_{Al}{}^2}{\{Al^{3+}\}^2\, E_{Ca}{}^3} = S_{Al/Ca} \quad (9.11)$$

or

$$\frac{\{Ca^{2+}\}^3\, [AlX_3]^2}{\{Al^{3+}\}^2\, [CaX_2]^3} \begin{array}{l} = \underline{S_{Al/Ca}} \\ = CEC \end{array} \quad (9.12)$$

where $S_{Al/Ca}$ is the selectivity coefficient for the Al–Ca exchange. Similar expressions hold for eqns. (9.8), (9.9), and (9.10).

The *cation exchange capacity*, of the soil is given by

$$CEC = [ALX_3] + [CaX_2] + [MgX_2] + [NaX] + KX] \quad (9.13)$$

where [ ] denotes the concentration of adsorbed ions in, say, milliequivalents per kilogram.

Expressed in terms of equivalent fractions of exchangeable ions, i.e. $E_{Ca} = [CaX]/CEC$ ... etc., the last equation becomes

$$E_{Al} + E_{Ca} + E_{Mg} + E_{Na} + E_K = 1 \quad (9.14)$$

The *base saturation* BS of the soil, defined as the sum of the equivalent fractions of all exchangeable ions except aluminium is then

$$BS = E_{Ca} + E_{Mg} + E_{Na} + N_K = 1 - E_{Al} \quad (9.15)$$

The four selectivity coefficients for cation exchange are not true thermodynamic constants since they vary from soil to soil and again have to be estimated from measurements of cation concentrations in the soil water and E_{Ca}/E_{Na}, E_{Mg}/E_{Na}, ... ratios on soil samples.

9.2.5 CALCULATIONS OF SOIL AND SOIL-WATER CHEMISTRY

Since there are 33 variables and only 24 equations, nine of the former have to be specified in advance from experimental data. MAGIC specifies the concentrations of Ca^{2+}, Mg^2, Na^+, K^+, $NH_4{}^+$, SO_4^{2-}, Cl^-, NO_3^-, which are then used to calculate

$[H^+]$ and hence the pH,

$[OH^-]$, $[HCO_3^-]$, $[CO_2^{2-}]$, $[OH^-]$ and hence the alkalinity of the soil water,

the concentrations of the 13 inorganic aluminium species and hence the total concentration of inorganic aluminium,

the base saturation and cation exchange capacity of the soil,

$p(CO_2)$, the partial pressure of CO_2 in the soil,

and $[CO_2(aq.)]$, the amount of dissolved CO_2 in soil water.

9.2.6 CALCULATION OF STREAM-WATER CHEMISTRY

When the percolate leaves the soil matrix and emerges as stream water, MAGIC assumes that no further exchange with the soil takes place, the concentrations of dissociated base cations and strong acid anions are unchanged, but when exposed to the atmosphere the excess CO_2 in the water degases. This causes a shift in the CO_3^{2-} – HCO_3^- equilibrium, a change in pH, and changes in concentrations of the aluminium species until equilibrium with solid gibbsite is restored. The relevant equations in the set are solved to compute the new values of pH and of alkalinity defined as

$$\begin{aligned} \text{alk} = {} & [HCO_3^-] + 2[CO_3^{2-}] + [OH^-] + [Al(OH)_4^-] \\ & -[H^+] - 3[Al^{3+}] - 2[AlOH^{2+}] - [Al(OH)_2^+] \end{aligned} \tag{9.16}$$

9.2.7 SOME APPLICATIONS OF THE MAGIC MODEL

The MAGIC model has been applied to the Allt a'Mharcaidh catchment by Jenkins *et al.* (1988). Firstly, the nitrate and ammonia uptake rates are optimized to match the output stream chemistry. Sulphate adsorption/desorption in the soil is linked to the concentration of dissolved sulphate by a Langmuir isotherm. The parameters controlling the cation behaviour, i.e. the weathering rates and the selectivity coefficients, which control cation exchange, together with the partial pressure of CO_2 in the soil, are optimized to match the stream chemistry and the measured base saturation of the soil. It turns out that the optimization process leads to the model having much higher concentrations of adsorbed sulphate and much lower weathering rates for the production of base cations than are deduced from field measurements. Nevertheless, the model has been used to compute the output stream chemistry for specified inputs of acid deposition over the 80-year period 1846–2026. A standard scenario assumes a linear increase in sulphate deposition from 1846 to 1916, then constant to

Table 9.1 Observed and simulated stream chemistry for the Allt a'Mharcaidh (in microequivalents per litre)

	1846 model simulated	1986 model simulated	1986 observed	2126 model simulated
Ca	21.4	37.5	37.1	43.8
Mg	28.6	29.6	29.9	29.7
Na	102.1	117.0	116.1	115.1
K	7.9	8.9	8.4	10.2
NH_4	0.0	2.0	—	2.2
SO_4	15.2	50.3	50.1	70.0
NO_3	0.0	2.2	2.1	2.2
Cl	111.3	111.3	111.3	111.3
Alk	33.5	31.6	33.0	17.2
H	1.6	1.7	2.0	2.8
pH	5.8	5.8	5.7	5.6

1936, followed by a linear rise to 1970, a 30 per cent fall by 1984, and thereafter remaining constant until AD 2126.

Application of this historical input data to Allt a'Mharcaidh indicates that the soils have retained a high buffering capacity since 1846, the alkalinity and pH of the streamwater having undergone little change as shown in Table 9.1. By 2126, if the deposition continues at its present rate, the alkalinity is predicted to drop to about 50 per cent of the present value and the pH to drop by about 0.2 units. The model's soil base saturation remains virtually unchanged over the whole 180-year period. The sulphate concentration in the stream increases threefold from 1846 to the present day but does not fall in response to the reduced deposition during 1970–84, and is predicted to increase gradually over the next few decades. These results stem largely from the model's high rates of sulphate adsorption on the soil; these would have to be reduced by an order of magnitude in order for the model to respond to the recent reductions in deposition.

Rather different results are obtained for the highly acidified Dargall Lane (Loch Dee) site where, for the same deposition scenario, the pH was computed to fall from 5.85 in 1846 to 5.35 in 1980, and thereafter to fall only slowly to 5.2 by 2060. There is little sulphate adsorption so the stream chemistry responds more strongly and more quickly to the changes in acid deposition.

MAGIC has also been applied to reconstruct the history, since 1846, of the stream chemistry at the Chon and Kelty catchments, both forested sites in an area of high deposition, 40 km north of Glasgow. Again, the

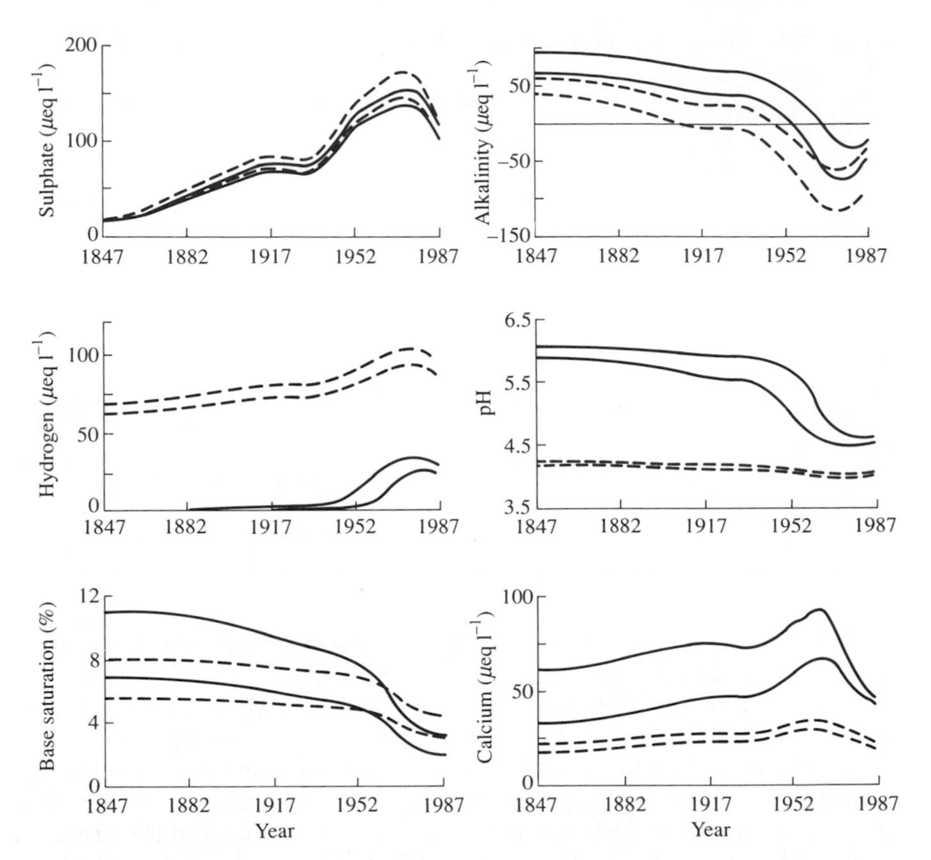

Fig. 9.1 Model simulations of stream chemistry and soil base saturation for the Chon (——) and Kelty (------) sites in central Scotland. Double lines represent the upper and lower ranges of expected behaviour. (From Whitehead *et al.* 1990, p. 436, by permission of the Cambridge University Press.)

model is optimized and tuned to give good agreement between the simulated and observed present-day stream and soil chemistries. Weathering rates for calcium and magnesium are considerably higher for the Chon simulation which accords with the discovery of a dolerite dyke — a strong source of these ions — within the catchment.

The model reconstructions depicted in Fig. 9.1 show Chon to have a very low H^+ concentration until 1950 when it increased sharply to reach a peak around 1970 and thereafter declined. The same trend holds for Kelty but with much higher background concentrations of H^+ ions derived from organic acids. The period of rapidly increasing acidity corresponds to the planting and growth of the forest reaching a peak as the forest canopy, which intercepts and enhances acid deposition, approaches complete closure. By 1980 the predicted stream acidity begins to decrease in response to falling deposition levels and to the decrease in uptake of base cations by the trees as the forest matures. This apparent recovery is in accord with the reconstructions from fossil diatom data (Battarbee *et al.* 1988, 1990).

It must be emphasized that, although these model reconstructions and predictions are very interesting and instructive they contain many uncertainties, some of which are inherent in the model formulation, which does not treat satisfactorily the aluminium chemistry, the production of base cations by weathering, and the sulphur processes in the soil. Sensitivity analyses, using a Monte Carlo technique, have revealed the simulated stream chemistry to be very sensitive to the calcium weathering rate and to the partial pressure of CO_2 in the soil and soil water but, in the absence of reliable measurements, it is difficult to know what values to choose.

In an attempt to remedy some of the most serious defects of MAGIC, recent modifications treat two of the soil processes, adsorption/desorption of sulphate and the production of base cations and strong-acid anions, as dynamic (time-dependent) rather than as equilibrium processes. Thus the changes in concentration of each ion are computed from equations of the form

$$\frac{dX}{dt} = F + W - QCq, \tag{9.17}$$

where X is the total amount of the ion per unit area of the catchment, F the deposition rate from the atmosphere, W the *net* uptake flux, including weathering, for the base cations, Q the flow rate of water through the catchment, C the concentration of free ions in the water, and q the ionic charge. After specifying the initial concentrations of the ionic species, subsequent changes are calculated from this equation as the computation proceeds.

9.3 The Birkenes model

This is a hydrochemical model, designed by Christophersen *et al.* (1982) to simulate *short-period* changes in stream flow and chemistry and to account for observed episodic and seasonal variations. The model has two soil reservoirs, one representing the shallow soils and the other the deeper soil layers. Storm-induced flow, acidic and rich in aluminium, is generated by rapid transfer of water in the upper reservoir, where the chemistry is assumed to be controlled by cation exchange and by aluminium dissolution and speciation in equilibrium with gibbsite. The less acidic and more calcium-rich base flow, stemming from the lower reservoir, is assumed to be controlled by mineral weathering and the adsorption/desorption of sulphate.

The input is the observed precipitation (rain or snow) corrected for evaporation and the percolating water from the two reservoirs is allowed to mix in independently determined proportions before leaving the system as stream water. The computed outputs are the concentrations of H^+, Al^{3+}, Ca^{2+}, Mg^{2+}, Na^+, SO_4^{2-}, HCO_3^- and organic anions and hence pH and alkalinity. The model is calibrated and tuned using past hydrological data to simulate the stream flow and stream chemistry of a particular catchment over a limited period and is then used to predict the flow and chemistry in response to changed inputs over different time periods and for other catchments.

The model reproduces the stream flow fairly well in most of the catchments to which it has been applied, although there is a marked tendency, like all models, to under predict the peak flows. However, the model is unable to reproduce the catchment's very damped response to large variations in the input of conservative (inert) species, such as chlorine (derived largely from sea salt) and the ^{18}O content of precipitation. When the model was applied to the Allt a'Mharcaidh catchment, it failed to predict the stream flow on an hourly time scale, especially during storm events when ^{16}O–^{18}O tracer studies show that much of the run-off is composed of pre-event water stored in the soil, the displacement of which, by so-called 'piston flow', causes a delayed response.

In attempts to overcome some of the model deficiencies, the formulation of water flow through the soil has been modified to ensure more rapid mixing and to allow for piston flow. The aluminium submodel is also being modified as measurements show that the assumption of equilibrium with gibbsite is generally invalid.

The MAGIC model has been used in combination with the Birkenes model, the former to simulate long-term changes in soil chemistry and the latter to simulate short-term changes in stream water chemistry for different acid deposition scenarios. The results indicate, for example, that a 30 per cent reduction in current deposition of excess sulphur would not be sufficient to halt soil acidification at a site like Birkenes. A 60 per cent reduction would produce significant improvements in stream water chemistry, but even a 90 per cent reduction would not guarantee successful restocking with trout.

A second approach utilizes a charge balance argument to estimate the minimum reduction in sulphate required to bring present H^+ and Al_i concentrations at high flow down to values acceptable for fish survival. The calculations contain many uncertainties but indicate that a 90 per cent reduction would not guarantee successful restocking at Birkenes although smaller reductions might make this possible in less acidified sites.

9.4 The Imperial College model

The BIRKENES and MAGIC models, with time resolutions of days to years, are unable to simulate the rapid changes in stream flow and stream chemistry that often follow events such as snow melt or rainstorms occurring on time scales of a few hours. Wheater *et al.* (1990) have therefore used their analyses of water flow on the plot, hillslope and catchment scales to develop an empirical hydrological model operating on hourly time steps. The incoming precipitation, adjusted for evaporation, is partitioned between reservoirs notionally representing the alpine, peaty-podzol, and peaty areas of the Allt a'Mharcaidh catchment and leads to the mixing of waters representing steady seepage, quick-flow, and ground-water components on their way to the stream.

As we have seen in Chapter 4, the hydrograph records show the stream flow during episodic events to consist of three components:

(1) large, short-period 'spikes' produced by quick flow through preferred channels and pipes, mainly in the upper peaty podzol;

(2) a 'humped' or 'whaleback' response, lasting sometimes for several days, produced by the downslope drainage of soil water through the lower peaty podzol;

(3) a base flow produced by a steady seepage of water through the soil profile and the peat near the valley bottom, most of which emerges after storage in the peat.

The water flow in the model is partitioned between reservoirs and along paths to simulate these three components, the partition coefficients being selected to achieve optimum fit to the recorded (hydrograph) flow over a 4-month period in the summer of 1988. Having calibrated the model in this way, achieving a regression coefficient $R^2 = 0.46$ between the simulated and observed data, it is then used to simulate the flow for two 5-month periods in each of two years with very different hydrological regimes. Agreement between the simulations and observations, albeit somewhat poorer than for the calibration run, is fairly good although the model fails to reproduce the highest observed peak flows — see Fig. 9.2.

Wheater *et al.* have extended the model in an attempt to simulate changes in stream water alkalinity during rapid-flow episodes, alkalinity being one of the few chemical parameters for which there are adequate data. Analysis by Harriman *et al.* (1990) of bi-weekly samples of the Allt a'Mharcaidh stream reveals a strong correlation between alkalinity and flow rate. For high flows the alkalinity is low and practically independent of the flow rate. For low flows the alkalinity is high and roughly inversely

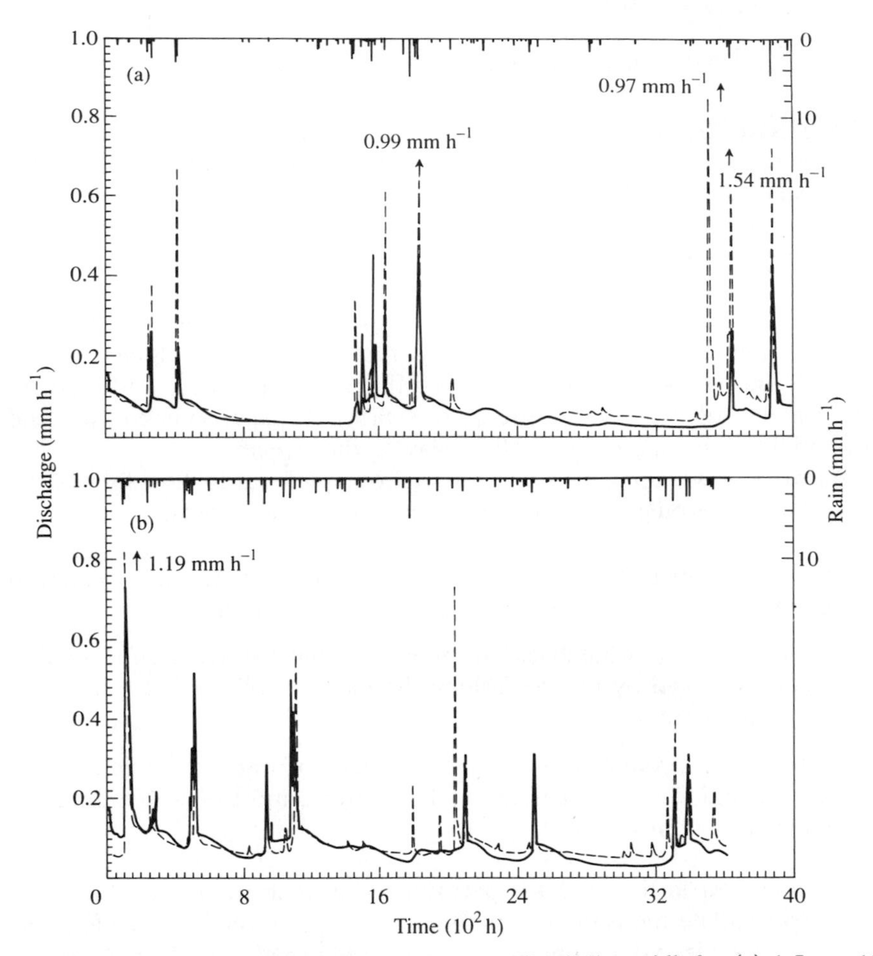

Fig. 9.2 Model simulations of stream flow at Allt a'Mharcaidh for (a) 1 June–15 November 1986, and (b) 3 June–31 October 1987: ——, simulated discharge; ------, observed discharge (From Wheater *et al.* 1990, p. 459, by permission of the Cambridge University Press.)

proportional to the flow rate. Since the soil water is generally acid, this indicates the presence of a source of high alkalinity, probably ground water. It is assumed that the alkalinity of the stream water results from simple conservative mixing of two or more components of the soil water, each having constant concentrations of the contributing ionic species.

The three components of soil-water flow identified in the hydrological model were allocated initial alkalinity values as follows:

(1) the highly alkaline base flow, alk = 90 μeq l^{-1};

(2) the slow downslope drainage component associated with the 'humped' response, alk = 90 μeq l^{-1};

(3) the quick-flow component through preferred paths associated with spikes on the hydrograph, alk = 0.

The mixing ratios between the three components are then chosen to obtain the best fit to Harriman's alkalinity measurements:

(1) during a snow-melt event on 18 March 1988 when the analysis led to the conclusion that more than 95 per cent of the stream flow originated from the quick-flow, low alkalinity component and less than 5 per cent from the high alkalinity, base flow;

(2) during a rainstorm on 22 July 1988 following a dry June, the quick-flow component through the dry soil again made the major (approximately 75 per cent) contribution to stream flow.

When this calibrated (tuned) model was applied to a longer series of data obtained in 1987 and 1988, the calculated changes in alkalinity over periods of 100–150 days agreed quite well with the observations, as shown in Fig. 9.3. Overall, the model, based on plot and hillslope experiments, has given some insight into the complexities of water transport and stream chemistry, and their interaction in a complete catchment, but in its present state it is unable to simulate the full stream-water chemistry, which may be much influenced by the efflux of pre-event water from the soil. There are, of course, considerable risks in applying a model based on observations from a limited area to a whole catchment, and even greater risks in applying such a model to a very different type of catchment.

9.5 General conclusions

In summary, much remains to be done to improve both the hydrology and the chemistry of models before they can simulate, much less predict, simultaneous changes of water flow and chemistry, on both short- and long-term scales. There are serious difficulties in representing adequately all the relevant chemical processes and the different water pathways, especially for a large, complex catchment having large variations in topography, slope, vegetation, soil structure, and composition. There is also an acute shortage of long-term data sets for calibration and validation of the models. Nevertheless, they give some indication of what changes to expect for different deposition scenarios and are proving useful in helping to identify the controlling and sensitive parameters and processes.

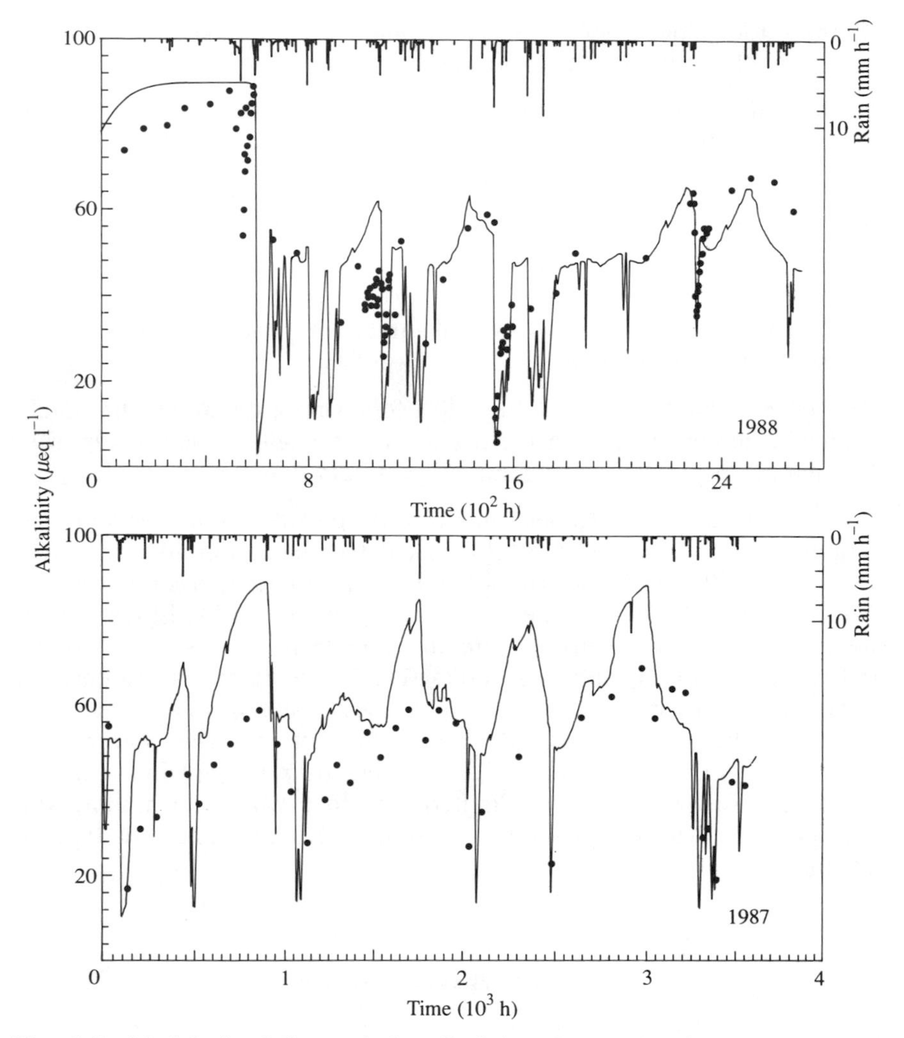

Fig. 9.3 Model simulations of the alkalinity changes in the stream at Allt a'Mharcaidh (——) and the observed values (·······) for one period in 1987 and one in 1988. (From Wheater *et al.* 1990, p. 464, by permission of the Cambridge University Press.)

9.6 References

Battarbee, R.W., *et al.* (1988). *Lake acidification in the United Kingdom, 1880–1986*. Ensis Publishing, London.

Battarbee, R.W., Mason, B.J., Renberg, I., and Talling, J.F. (eds.) (1990). *Palaeolimnology and lake acidification*. The Royal Society, London.

Christophersen, N., Seip, H.M., and Wright, R.F. (1982). *Water Resources Research*, **18**, 977–96.

Cosby, B.J., *et al.* (1985). *Water Resources Research*, **21**, 51–63.
Harriman, R., *et al.* (1990). In *The surface waters acidification programme*, (ed. B.J. Mason), pp. 31–45. Cambridge University Press.
Jenkins, A., *et al.* (1988). *Water, Air*, and *Soil Pollution*, **40**, 275–91.
Wheater, H.S., *et al.* (1990). In *The surface waters acidification programme*, (ed. B.J. Mason), pp. 455–66. Cambridge University Press.
Whitehead, P.G., Jenkins, A., and Cosby, B.J. (1990). In *The surface waters acidification programme*, (ed. B.J. Mason), pp. 431–43. Cambridge University Press.

FURTHER READING

Christophersen, N., *et al.* (1990). In *The surface waters acidification programme*, (ed. B.J. Mason), pp. 445–52. Cambridge University Press.
B.J. Mason (ed.) (1990). *The surface waters acidification programme*, pp. 431–505. Cambridge University Press.
J. Kamari (ed.) (1990). *Impact models to assess regional acidification*. Kluwer, Amsterdam.
Whitehead, P.G., *et al.* (1988). *Journal of Hydrology*, **101**, 191–212.

10
Manipulation experiments on catchments

10.1 Introduction

Increasingly conclusive evidence that acid deposition results in the acidification of surface waters and the loss of fish populations has led to international pressure and action to reduce the emissions of sulphur and nitrogen oxides on the assumption that this will result in at least a partial recovery of water quality and fish populations. In order to test this premise, manipulation experiments have been carried out on whole catchments in Norway in which the acid loading is changed in a controlled manner.

The RAIN project (Reversing Acidification in Norway), described by Wright and Henriksen (1990), comprises two experiments: artifical acidification of two pristine catchments in western Norway (Sogndal site), and the exclusion of ambient acid deposition, by means of a roof, from a strongly acidified catchment in southernmost Norway (Risdalsheia site). The experiments have now been conducted for 7 years, since 1984.

10.2 The Sogndal acidification experiment

The Sogndal site is a pristine but sensitive area in western Norway receiving only weakly acidic precipitation of pH 4.8. It is located 900 m above sea level on gneissic bedrock, with patchy, thin, and poorly developed soils having pH 4.5–5.5, and alpine vegetation. One catchment, SOG2, is treated with sulphuric acid, another, SOG4, with sulphuric and nitric acid, and two much larger catchments, SOG1 and SOG3, serve as untreated controls. The treated catchments receive identical loadings $[H^+] = 100$ meq m^{-2} yr,$^{-1}$, typical of natural levels in southern Norway, an acid–lake-water mixture being applied to the snowpack at 2 mm h^{-1}.

The volume and chemical composition of natural rain and snow are measured weekly about 3 km from the experimental sites. Stream flow is measured continuously at the outlets of three of the catchments. Stream samples are chemically analysed weekly but almost daily during snow-melt. During the period of acid treatment, samples are taken every 2 hours and daily for 5 days afterwards.

Each application of acid produces a short-term pulse of acid, aluminium-rich run-off. Recovery between episodes is generally rapid with full recovery in the first year but slower and less complete in later years. Treatment with H_2SO_4 in SOG2 catchment increased the mean $[SO_4]$ in the stream from 20–50 to 50–60 μeq l^{-1}, while the H_2SO_4 + HNO_3 treatment in SOG4 produced SO_4 levels of 35–45 μeq l^{-1}. These levels are only about 50 per cent of the steady-state values calculated on the assumption that the SO_4 flux in the run-off is equal to the SO_4 deposition. Thus much of the incoming SO_4 (about 80 per cent over 4 years) was stored in the catchment, mainly by adsorption on the soil to form a readily available pool. The addition of HNO_3 caused only minor increases in the nitrate run-off of SOG4 until 1989, when signs of nitrate leaching began to show. The increased concentrations of SO_4^{2-} in run-off from both treated catchments are compensated about 50 per cent by increased concentrations of Ca^{2+} and Mg^{2+} released from the soil by cation exchange with the incoming H^+ ions, and about 50 per cent by decreased alkalinity, defined as $[HCO_3^- + 2CO_3^{2-} + OH^- - H^+]$. After 3 years of treatment, alkalinity was strongly negative, bicarbonate levels negligible, and aluminium levels high, making the run-off lethal to salmon fry.

Concentrations of organic aluminium are very low but acid treatment results in a sharp increase in inorganic aluminium as the acid dissolves aluminium-hydroxide precipitates on the stream bed. However, the aluminium solubility is not consistent with equilibrium with respect to a single phase of aluminium hydroxide (e.g. gibbsite).

The net fluxes of Ca^{2+} + Mg^{2+} from the treated catchments were about 50 per cent higher than in the controls, caused either by increased weathering or, more likely, by depletion of exchangeable base cations in the soil. If the latter were entirely responsible, then the depletion over 4 years of acid treatment was about 2 per cent. This suggests that a significant decrease in soil base saturation could occur after a very few decades of acid deposition, such as occur in southern Norway.

10.3 The Risdalsheia acid exclusion experiment

The Risdalsheia site is in an acid-sensitive area in southern Norway, with a high total sulphate deposition of 100 meq m^{-2} yr^{-1}, the mean pH of precipitation being 4.2. It is located 300 m above sea level on exposed granitic bedrock with thin, organic-rich soil (average depth 10–15 cm, pH 3.9–4.5) and a sparse cover of pine and birch.

Acid deposition is excluded from the KIM sub-catchment of area 860 m^2 by a 1200 m^2 transparent roof. Incoming precipitation is collected from the roof, pumped through filters and an ion-exchange system. Sea-water salts are added back and the clean precipitation is then applied beneath the roof above the vegetation canopy by a sprinkler at 2 mm h^{-1}. An adjacent catchment (EGIL) of 400 m^2 is also covered, but receives the untreated acid precipitation by an identical sprinkler. A third uncovered catchment (ROLF) acts as a control. Treatment began in June 1984. Precipitation volume and chemistry are measured at weekly intervals. Stream flow is measured continuously and sampled for detailed chemical analysis at least once a week.

Exclusion of wet acid deposition at KIM caused the concentration of sulphate in the run-off to decrease sharply at first, and then more slowly, so that in 1990 it was still 40–50 μeq l^{-1} or about half that in run-off from the uncovered catchment. So, after 6 years of wet acid exclusion, equilibrium has not yet been reached between output and the drastically reduced input. This suggests a steady leaching of sulphate stored in the soil, but the rate is not easy to determine because of uncertainty in the input from dry deposition. However, the total amount of water-soluble and adsorbed sulphate in the soil prior to acid exclusion is far from sufficient to account for the continued high concentrations in the run-off and points to the existence of a large reservoir of organic sulphate in the soil.

The alkalinity of the run-off has increased from −100 to −30 μeq l^{-1} since acid exclusion and reflects both the decrease of H^+ and aluminium ions, and the increased dissociation of organic acids.

The pH of the run-off has increased much less than expected, the current average value of 4.3 being only a little higher than that (pH 4.0) for the control catchments. This is because the water is highly buffered by organic anions produced by the dissociation of organic acids, the concentrations having steadily increased during the exclusion experiment to around 40–50 μeq l^{-1} in 1990. The run-off remains highly toxic to brown trout, probably because the very low concentrations of calcium (<0.5 mg l^{-1}) are unable to protect them against the prevailing concentrations of aluminium. However, many lakes in southern Norway have very low levels of total organic carbon and have higher concentrations of base cations, so that reduced acid deposition should lead to more rapid recovery of pH than for Risdalsheia.

The latter experiment is expected to continue for several more years in order to determine the long-term recovery rate of this particularly unfavoured and vulnerable catchment.

10.4 The Loch Fleet liming project

The Swedish Fisheries Board has shown that treating acid lakes, containing only low concentrations of neutralizing calcium, with powdered limes-

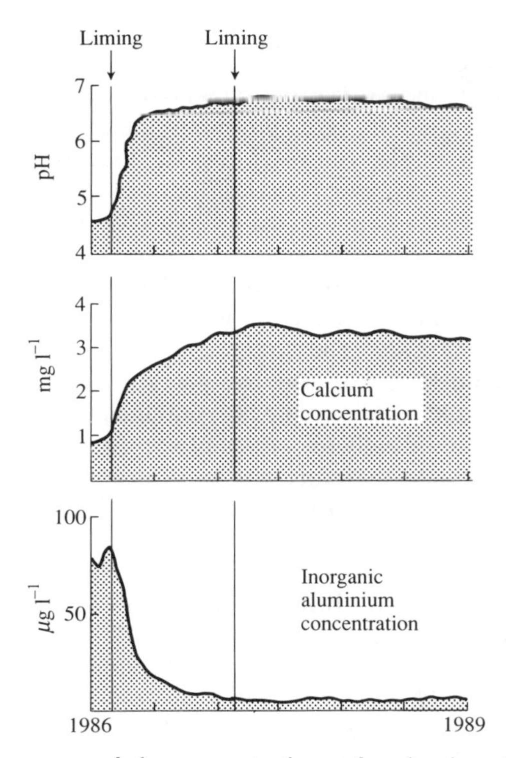

Fig. 10.1 Time response of the concentrations of major ions in the tributary water at Loch Fleet following liming of the surrounding catchment area. (From CEGB *et al.* 1989.)

tone can quickly restore water quality and, in due course, fisheries. The limestone is usually dispersed over the lake from a boat or, in remote areas, from a helicopter.

The Loch Fleet project was established by the Central Electricity Generating Board, South of Scotland Electricity Board, North of Scotland Hydro-Electric Board, and British Coal to investigate a possible alternative method, namely liming the surrounding land rather than the lake, in the belief that this would be cheaper and have longer lasting effects.

The loch, located in Galloway in south-west Scotland, underwent acute acidification in the early 1970s and became fishless in 1975. The 5-year study was designed to discover the response of the lake and its catchment to various forms of lime treatment. Over 300 tonnes of limestone were applied to various parts of the catchment in April 1986 with further applications in other parts in 1987. As Fig. 10.1 shows, the pH and the calcium and inorganic aluminium concentrations responded beneficially in a very short time and these improvements have persisted. Brown trout were reintroduced in May 1987. These thrived and spread throughout the lake and downstream.

Experience from the project indicates that liming should be applied in excess of 5 tonnes per hectare, preferably in areas such as bogs which ensure that much of the run-off water is intercepted. Best results are obtained with fine powder or pellets of limestone or chalk. The operation is not cheap, the cost being at least £100–£350 per hectare, for lasting effects.

It appears that reduction of emissions of SO_2 and NO_x will need to be complemented by liming strongly acidified catchments if the water quality is to be restored quickly. The cost will be small compared with that of removing SO_2 and NO_x from the flue gases of power stations.

On the other hand, it is possible that liming may have some deleterious long-term effects, especially on trees and vegetation that are well adapted to acid soils. It should therefore not be regarded as a universal panacea without further research and monitoring.

10.5 References

Wright, R.F. and Henriksen, A. (1990). In *The surface waters acidification programme*, (ed. B.J. Mason), pp. 161–6. Cambridge University Press.

Central Electricity Generating Board (CEGB), *et al.* (1989). *New life for acid waters; the Loch Fleet Project*. CEGB, London.

Appendix: Main conclusions of the Surface Waters Acidification Programme

1. Acidified lakes and streams without, or with impoverished, fish populations occur mainly in areas that receive high levels of acid deposition from the atmosphere and have soils derived from granite or other rocks of similar composition that are resistant to weathering and low in exchangeable elements such as calcium and magnesium. Catchments with thin soils are particularly sensitive with respect to the rate and extent of acidification.

2. Examination of the remains of diatoms and other biological material in lake sediments laid down over centuries has established that many lakes in southern Norway and Sweden and in the UK have undergone progressive acidification from *circa* 1850 until very recently. The magnitude of this acidification is appreciably greater than any that has occurred in the past 10 000 years and has marched in parallel with accelerated industrial development, as indicated by increases in several trace pollutants in the sediments. These changes and the extent of inferred acidification are geographically correlated with the intensity of acid deposition and with the geochemical status of the catchment.

3. For a given input of acid deposition, the degree of acidification of lakes and streams is largely determined by the structure and chemistry of the mineral and organic soils, and the pathways that the incoming rainwater takes through the soil. These factors determine both the nature and duration of the many chemical and biological reactions that influence the final quality of the water that emerges in the streams.

4. The evidence points convincingly to atmospheric deposition, largely of acidifying compounds of sulphur and to a lesser extent of nitrogen, as the main cause of acidification. However, forests may enhance acidification by acting as efficient filters and collectors of acid from the atmosphere in polluted areas, and by taking up metal cations, and the acidification of some lakes may be attributed to changes in land use or agricultural practice.

5. There is evidence that, in the past decade, that there has been a significant decrease in the acidity of rain and snow as a result of reduced emissions of sulphur dioxide, and that this is reflected in a small decline in the acidity and sulphate content of some lakes. However, there are signs,

especially in Norwegian lakes, that the effects of reduced concentrations of sulphate are being partially offset by increases in nitrate.

6. Fish populations, especially of salmon and trout, cannot survive in lakes and streams if the pH of the water remains below a critical level of about pH 5 for long (depending on the species and age of the fish and the chemical composition of the water). The fish are killed by the action of increased acidity and of inorganic forms of aluminium leached out of the soil by the acidified water. The effects of aluminium are ameliorated by the presence of organic acids (e.g. from peat) that complex the aluminium and render it less toxic to fish, and possibly if calcium is present in sufficiently high concentration. However, in regression analyses based on a survey of over 1000 lakes in southern Norway in which 14 variables in the regressions were studied, most of the variance in fishery status could be accounted for by pH, inorganic aluminium, and altitude.

7. Fish are very vulnerable to the short, sharp episodes of high acidity and aluminium that occur in streams following heavy rains or snow melt. In these episodes much of the water flows through acid soils where it is enriched in available aluminium but spends relatively little or no time in the deeper layers where it would be neutralized.

8. From carefully controlled laboratory experiments and intensive field studies, it is now possible to relate fish survival to the concentrations of acid, aluminium, and calcium in the water and to estimate the likely toxic effects of acidic episodes of differing severity, frequency, and duration. The detailed mechanisms of fish death are complex. Some forms of aluminium and H^+ ions inhibit sodium uptake and increase sodium loss, so reducing body sodium content leading, eventually, to circulatory failure. The deleterious effects of inorganic aluminium can be largely counteracted if calcium is present in sufficient concentrations.

9. Acidification and release of aluminium also leads to changes in the populations of micro-organisms, lower plants, and aquatic invertebrates. The effects of such changes in the ecosystem can include the availability of food for some life stages of brown trout and other fish.

10. The possibilities of the recovery of streams and lakes depend on the long-term balance between the catchment input and output of cations, such as Ca^{2+} and Mg^{2+}, that are exchangeable for H^+ ions. The main input of these cations in the affected parts of Scandinavia and the UK is normally by chemical weathering; the supply by atmospheric deposition is much less than that of acidifying substances. The direct cause of acidification of lakes and streams is the excess of anions of strong acids, as sulphate reduction and denitrification play only a minor role in most of the affected ecosystems. The long-term resistance of a catchment area is therefore closely related to the release of cations such as Ca^{2+}, Mg^{2+}, Na^+, and K^+. In regions covered by the last glaciation, these cations are produced mainly by weathering of minerals. Considerable progress has

been made within SWAP in the determination of weathering rates as a function of mineral species, particle size, pH, production of organic ligands in the ecosystem, and the history of the soil. Estimates by different methods agree in most cases within a factor of 2 or 3. It appears that even a reduction by 60 per cent of acid deposition would not be enough to create steady-state conditions suitable for fish in those areas that are most strongly acidified.

11. The rate at which streams and lakes will recover in response to reduced emission and deposition of acidic substances will also depend on such factors as the residence time of water in ground water and lakes, on the release of sulphate from earlier deposition, which is retained in the soil, and on changes in the land use within the catchment. In thin soils with little storage of sulphur compounds, recovery may be quite rapid. In deeper soils containing large accumulated stores of sulphur compounds, it may take several years or even decades for this to be leached out and recovery may be much slower. Recovery or restoration may be aided by liming the catchment, but this may have undesirable effects such as increased nitrification.

12. There is evidence of increased nitrate deposition but this has been only partly reflected by the increase in its concentration in surface waters, mainly because of uptake by vegetation. As the system has limited storage capacity, an additional burden of acidification could develop over years.

Index